Guidelines for Applying Protected Area Management Categories

IUCN

Founded in 1948, IUCN (International Union for Conservation of Nature) brings together States, government agencies and a diverse range of non-governmental organizations in a unique world partnership: over 1000 members in all, spread across some 160 countries. As a Union, IUCN seeks to influence, encourage and assist societies throughout the world to conserve the integrity and diversity of nature and to ensure that any use of natural resources is equitable and ecologically sustainable. IUCN builds on the strengths of its members, networks and partners to enhance their capacity and to support global alliances to safeguard natural resources at local, regional and global levels.

Website: www.iucn.org

The World Commission on Protected Areas (WCPA)

The World Commission on Protected Areas (WCPA) is the world's leading network of protected area managers and specialists, with over 1,300 members in 140 countries. WCPA is one of the six voluntary Commissions of IUCN and is administered by the Programme on Protected Areas at IUCN's headquarters in Gland, Switzerland. WCPA's mission is to promote the establishment and effective management of a worldwide representative network of terrestrial and marine protected areas, as an integral contribution to the IUCN mission.

Website: www.iucn.org/themes/wcpa

Regional Council for the Environment of Junta de Andalucía

The Regional Council for the Environment of Junta de Andalucía is the agency of the regional government of Andalucía responsible for the conservation of nature, the application of environmental regulations and policies on the use and management of natural resources, the declaration and management of protected areas, as well as the definition, development and implementation of climate change mitigation and adaptation strategy and policies.

Fundación Biodiversidad

The Fundación Biodiversidad (Biodiversity Foundation) is a non-profit organization established in 1998 following the commitments undertaken by Spain after the ratification of the Convention on Biological Diversity. It carries out activities in the field of conservation, study, and sustainable use of biodiversity, as well as in international development cooperation. Through International Cooperation, the Fundación Biodiversidad manages to unite efforts and create synergies, as well as to promote collaboration with national and international organizations, institutions and programmes.

Guidelines for Applying Protected Area Management Categories

Edited by Nigel Dudley

Published by:	IUCN, Gland, Switzerland
Copyright:	© 2008 International Union for Conservation of Nature and Natural Resources
	Reproduction of this publication for educational or other non-commercial purposes is authorized without prior written permission from the copyright holder provided the source is fully acknowledged.
	Reproduction of this publication for resale or other commercial purposes is prohibited without prior written permission of the copyright holder.
Citation:	Dudley, N. (Editor) (2008). *Guidelines for Applying Protected Area Management Categories.* Gland, Switzerland: IUCN. x + 86pp.
ISBN:	978-2-8317-1086-0
Cover photos:	Front: Discussion with local communities near Morondava, Madagascar about zoning in a proposed protected area to conserve rare baobab tree species © Nigel Dudley Back: New Caledonia © Dan Laffoley
Layout by:	Bookcraft Ltd, Stroud, UK
Produced by:	IUCN Publications Services
Printed by:	Page Bros, Norwich, UK
Available from:	IUCN (International Union for Conservation of Nature) Publications Services Rue Mauverney 28 1196 Gland Switzerland Tel +41 22 999 0000 Fax +41 22 999 0020 books@iucn.org www.iucn.org/publications

A catalogue of IUCN publications is also available.

The text of this book is printed on Greencoat Velvet 100gsm (recycled, FSC).

Contents

Foreword

Protected areas remain the fundamental building blocks of virtually all national and international conservation strategies, supported by governments and international institutions such as the Convention on Biological Diversity. They provide the core of efforts to protect the world's threatened species and are increasingly recognised as essential providers of ecosystem services and biological resources; key components in climate change mitigation strategies; and in some cases also vehicles for protecting threatened human communities or sites of great cultural and spiritual value. Covering almost 12 percent of the world's land surface, the global protected area system represents a unique commitment to the future; a beacon of hope in what sometimes seems to be a depressing slide into environmental and social decline.

Protected areas are by no means uniform entities however; they have a wide range of management aims and are governed by many different stakeholders. At one extreme a few sites are so important and so fragile that no-one is allowed inside, whereas other protected areas encompass traditional, inhabited landscapes and seascapes where human actions have shaped cultural landscapes with high biodiversity. Some sites are owned and managed by governments, others by private individuals, companies, communities and faith groups. We are coming to realize that there is a far wider variety of governance than we had hitherto assumed.

The IUCN protected area management categories are a global framework, recognised by the Convention on Biological Diversity, for categorizing the variety of protected area management types. Squeezing the almost infinite array of approaches into six categories can never be more than an approximation. But the depth of interest and the passion of the debate surrounding the revision of these categories show that for many conservationists, and others, they represent a critical overarching framework that helps to shape the management and the priorities of protected areas around the world.

We have not rushed this revision. It began with a two-year consultative research project that reported to the World Conservation Congress in Bangkok in 2004, resulting in a resolution calling for the production of the guidelines presented in this book. In the years since, IUCN has consulted with a huge number of its members in special meetings, conferences, electronic debates and through what sometimes seemed like an endless correspondence. We are well aware that the results are not perfect – an impossible task. But we believe the interpretation of the protected area definition and categories presented here represents the opinion of the large majority of IUCN members. Importantly, they are complemented by the IUCN governance types, demonstrating the importance that the Union is giving to issues of governance.

In the years to come we will be working to promote the category system, to translate the guidelines into more languages and to make sure they are applied effectively, in order to maximize the potential of the global protected area system in perpetuity.

Acknowledgements

The revision of the IUCN guidelines has followed a long and exhaustive process of consultation within IUCN. We are deeply grateful to members of IUCN, the IUCN World Commission on Protected Areas and the Task Force on Categories for help in developing and agreeing the final text. This publication is the result of this revision and it has been made possible due to the generous financial contribution from Fundación Biodiversidad of Spain. Fundación Biodiversidad (Biodiversity Foundation) is a non-profit foundation established in 1998 following the commitments undertaken by Spain after the ratification of the Convention on Biological Diversity. It carries out activities in the field of conservation, study, and sustainable use of biodiversity, as well as in international cooperation for development. Through International Cooperation, the Fundación Biodiversidad manages to unite efforts and create synergies, as well as to promote collaboration with national and international organizations, institutions and programmes.

First, we thank the many people who commented on the *Speaking a Common Language* project, resulting in a final report written by Kevin Bishop, Nigel Dudley, Adrian Phillips and Sue Stolton, which formed the background research leading to the revision of the categories. A full acknowledgements list is included in the report from this project, but more recently we should single out Natalia Danilina, WCPA Vice-Chair for North Eurasia, for arranging translation of the whole report into Russian.

Next, grateful thanks are extended to all the people who wrote commissioned or independent papers on application of the categories and suggestions for revised text. These include: Robin Abell, José Antonio Atauri, Christian Barthod, Charles Besancon, Harry Biggs, Luigi Boitani, Grazia Borrini-Feyerabend, Peter Bridgewater, Jessica Brown, Phillip Bubb, Neil Burgess, José Courrau, Roger Crofts, Nick Davidson, Jon Day, Phillip Deardon, Benita Dillon, Charlie Falzon, Lucy Fish, Pete Frost, Roberto Gambino, John Gordon, Craig Groves, David Harmon, Marc Hockings, Sachin Kapila, Cyril Kormos, Ashish Kothari, Dan Laffoley, Harvey Locke, Stephanie Mansourian, Josep-Maria Mallarach, Claudio Maretti, Carole Martinez, Kenton Miller, Brent Mitchell, John Morrison, C. Niel, Gonzalo Oviedo, Jeffrey Parrish, Andrew Parsons, Marc Patry, Jean-Marie Petit, Adrian Phillips, Kent Redford, Liesbeth Renders, Carlo Rondinini, Deborah Bird Rose, Fausto Sarmiento, David Sheppard, Daniela Talamo, Daniel Vallauri, Bas Verschuuren, John Waugh and Bobby Wishitemi. Funding for the production of some of these papers came from BP and we are very grateful for their support.

A critical part of this revision process was the implementation of the IUCN Categories Summit, held in Almería, Spain (7–11 May, 2007). The Categories Summit was organized and implemented with financial and institutional support from Junta de Andalucía, Fundación Biodiversidad and the IUCN Centre for Mediterranean Cooperation. The Regional Council for the Environment of Junta de Andalucía provided logistical and technical support during the Summit, in the form of case studies and field activities, that substantially contributed to its success. The Regional Council for the Environment of Junta de Andalucía is the agency of the regional government of Andalucía responsible for the conservation of nature, the application of environmental regulations and policies on the use and management of natural resources, the declaration and management of protected areas, as well as the definition, development and implementation of climate change mitigation and adaptation strategies and policies.

A large number of people gave up a week of their time to discuss the revision of the categories during the IUCN Categories Summit. Particular thanks are due to the following experts who participated: Tarek Abulhawa, Andrés Alcantara, Germán Andrade, Alexandru Andrasanu, Suade Arancli, Margarita Astralaga, José Antonio Altauri, Jim Barborak, Brad Barr, Christian Barthod, Louis Bélanger, Charles Besancon, Ben Böer, Grazia Borrini-Feyerabend, Peter Bridgewater, Tom Brooks, Jessica Brown, Susana Calvo Roy, Sonia Castenáda, Carles Castell Puig, Miguel Castroviejo Bolivar, Peter Cochrane, Peter Coombes, José Courrau, Botella Coves, Roger Crofts, Marti Domènech I Montagut, Marc Dourojeanni, Holly Dublin, Nigel Dudley, Abdellah El Mastour, Ernest Enkerlin Hoeflicj, Reinaldo Estrada, Jordi Falgarona-Bosch, Antonio Fernández de Tejada González, Georg Frank, Roberto Gambino, Javier Garat, Sarah Gindre, Craig Groves, José Romero Guirado, Manuel Francisco Gutiérrez, Heo Hag-Young, Marc Hockings, Rolf Hogan, Bruce Jeffries, Vicente Jurado, Ali Kaka, Sachin Kapila, Seong-II Kim, Cyril Kormos, Meike Kretschmar, Zoltan Kun, Dan Laffoley, Kari Lahti, Maximo Liberman Cruz, Harvey Locke, Axel Loehken, Arturo Lopez, Elena López de Montenegro, Nik Lopoukhine, Ibanez Luque, Maher Mahjoub, Josep Maria Mallarach, Moses Mapesa, Claudio Maretti, Vance Martin, María Teresa Martín Crespo, Carole Martinez, Baldomero Martinez, Julia Marton-Lefèvre, Mehrasa Pehrdadi, Rosa Mendoza Castellón, Kenton Miller, Susan Miller, Carmen Miranda, Fernando Molina, Sophie Moreau, Gérard Moulinas, Marta Múgica, Eduard Müller, Anread Müseler, Olav Nord-Varhaug, Juan Carlos Orella, Gonzalo Oviedo, Ana Pena, Milagros Pérez Villalba, Christine Pergent-Martini, Rosario Pintos Martin, Anabelle Plantilla, Francisco Quiros, Mohammed Rafiq, Tamica Rahming, Anitry Ny Aina Ratsifandrihamanana, Kent Redford, Manuel Rodriguez de Los Santos, Pedro Rosabal, Juan Carlos Rubio Garcia, Alberto Salas, Francisco Sanchez, Ana Elena Sánchez de

Dios, José Luis Sánchez Morales, Mohammed Seghir Melouhi, Peter Shadie, David Sheppard, Sue Stolton, Gustavo Suárez de Freitas, Daniela Talamo, Tony Turner, Rauno Väisänen, Tafe Veselaj, Nestor Windevoxhel and Stephen Woodley.

In addition, regional meetings were held to discuss the categories at the 2nd ASEAN Heritage Parks Conference and 4th Regional Conference on Protected Areas in South East Asia in Sabah, Malaysia; in association with the UNEP World Conservation Monitoring Centre in Nairobi, Kenya; at the Second Latin American Parks Congress in Bariloche, Argentina and at the WCPA European Meeting in Barcelona, Spain. We are grateful to the organizers, including Christi Nozawa, Anabelle Plantilla, Geoffrey Howard, Sue Stolton, Carmen Miranda and Roger Crofts. We are also grateful to all the people who took part in the workshops and whose ideas contributed to the final guidelines.

Meetings also took place at the International Council on Mining and Metals and the International Petroleum Environmental Conservation Association, both in London, and at a special meeting of industry representatives with IUCN in Gland, Switzerland, and we thank the organizers of these events.

Many people commented on the protected area definition, the whole guidelines or part of the guidelines and many more contributed to the e-debate. Amongst those who sent written comments or took part in or organized meetings were, in addition to people already listed above: Mike Appleton, Alberto Arroyo, Andrea Athanus, Tim Badman, John Benson, Juan Bezaury, Stuart Blanch, Andrer Bouchard, José Briha, Kenneth Buk, Eduardo Carqueijeiro, Brian Child, Thomas Cobb, Nick Conner, Marina Cracco, Adrian Davey, Fekadu Desta, Jean Pierre d'Huart, Paul Eagles, Joerg Elbers, Neil Ellis, Penny Figgis, Frauke Fisher, James Fitzsimmons, Gustavo Fonseca, Alistair Gammell, George Gann, Brian Gilligan, Fernando Ghersi, Hugh Govan, Mary Grealey, Michael Green, Larry Hamilton, Elery Hamilton Smith, Alan Hemmings, John Hough, Pierre Hunkeler, Glen Hvengaard, Tilman Jaeger, Jan Jenik, Graeme Kelleher, Richard Kenchington, Saskia de Koning, Linda Krueger, Barbara Lausche, Richard Leakey, Mary Kay LeFevour, Li Lifeng, Heather MacKay, Brendan Mackey, Dave MacKinnon, Vinod Mathur, Nigel Maxted, Jeffrey McNeely, Mariana Mesquita, Paul Mitchell, Russ Mittermeier, Geoff Mosley, Fulori Nainoca, Juan Oltremari, Sarah Otterstrom, Thymio Papayanis, Jamie Pittock, Sarah Pizzey, Dave Pritchard, Allen Putney, Joanna Robertson, Jaime Rovira, Tove Maria Ryding, Heliodoro Sánchez, Andrej Sovinc, Rania Spyropoulou, Erica Stanciu, David Stroud, Surin Suksawan, Martin Taylor, Djafarou Tiomoko, Joseph Ronald Toussaint, Frank Vorhies, Daan Vreugdenhil, Haydn Washington, Sue Wells, Rob Wild, Graeme Worboys, Eugene Wystorbets and Edgard Yerena. Many people sent in collective responses, reflecting a number of colleagues or an institution or NGO.

David Sheppard, Pedro Rosabal, Kari Lahti and Tim Badman, from the IUCN Programme on Protected Areas (PPA), have provided technical input and policy guidance throughout this process; Delwyn Dupuis, Anne Erb and Joanna Erfani (PPA) have also provided much-needed administrative assistance and support from the IUCN Headquarters in Gland. Nik Lopoukhine, Chair of WCPA, has been constant in his support for this process, as have the members of the WCPA Steering Committee. In particular Trevor Sandwith, Roger Crofts and Marc Hockings all gave detailed readings of the entire text and Grazia Borrini-Feyerabend and Ashish Kothari have commented on numerous versions of the section on governance. Technical and policy advice from Gonzalo Oviedo, IUCN Senior Adviser on Social Policy, was fundamental in relation to governance and indigenous peoples issues.

Peter Cochrane and Sarah Pizzey of Parks Australia arranged and supported a lengthy trip to five states in Australia to discuss the categories with dozens of protected area professionals both in meetings and in the field. This input added greatly to our understanding of the challenges and opportunities in setting new guidelines and allowed us to test out ideas.

Work on category Ib has been driven by the Wilderness Task Force chaired by Vance Martin, with the lead on the categories being taken by Cyril Kormos. The position on IUCN category V has been developed further through two meetings of the special task force dedicated to landscape approaches, generously funded by the Catalan government and by a consortium of conservation agencies in the UK: Natural England, Scottish Natural Heritage and the Countryside Council for Wales. Jessica Brown chairs the task force and organized the meetings, with help from respectively Jordi Falgarone and Andy Brown. The position on category VI has been developed through the work of a new Category VI Task Force chaired by Claudio Maretti and at a meeting as part of the Latin America and Caribbean Parks Congress at Bariloche, Argentina.

Introduction

The following guidelines are offered to help in application of the IUCN protected area management categories, which classify protected areas according to their management objectives. The categories are recognised by international bodies such as the United Nations and by many national governments as the global standard for defining and recording protected areas and as such are increasingly being incorporated into government legislation. For example, the CBD *Programme of Work on Protected Areas* "*recognizes the value of a single international classification system for protected areas and the benefit of providing information that is comparable across countries and regions and therefore welcomes the ongoing efforts of the IUCN World Commission on Protected Areas to refine the IUCN system of categories … *"

The guidelines provide as much clarity as possible regarding the meaning and application of the categories. They describe the definition and the categories and discuss application in particular biomes and management approaches.

The original intent of the IUCN Protected Area Management Categories system was to create a common understanding of protected areas, both within and between countries. This is set out in the introduction to the Guidelines by the then Chair of CNPPA (Commission on National Parks and Protected Areas, now known as the World Commission on Protected Areas), P.H.C. (Bing) Lucas who wrote: "*These guidelines have a special significance as they are intended for everyone involved in protected areas, providing a common language by which managers, planners, researchers, politicians and citizens groups in all countries can exchange information and views*" (IUCN 1994).

As noted by Phillips (2007) the 1994 Guidelines also aimed to: "*reduce the confusion around the use of many different terms to describe protected areas; provide international standards for global and regional accounting and comparisons between countries, using a common framework for the collection, handling and dissemination of protected areas data; and generally to improve communication and understanding between all those engaged in conservation*".

This use of the protected area categories as a vehicle for "speaking a common language" has considerably broadened since the adoption of the guidelines in 1994. In particular, there have been a number of applications of the categories system in policy at a range of levels: international, regional and national. The current guidelines thus cover a wider range of issues and give more detail than the 1994 version. They will, as necessary, be supplemented by more detailed guidance to individual categories, application in particular biomes and other specialized areas. Following extensive consultation within IUCN and with its members, a number of additional changes have been made since 1994, including to the definition of a protected area and to some of the categories.

Should "protected area" be an inclusive or exclusive term?

One fundamental question relating to the definition and categories of protected areas is whether the word "protected area" should be a general term that can embrace a very wide range of land and water management types that *incidentally* have some value for biodiversity and landscape conservation, or instead be a more precise term that describes a particular form of management system especially *aimed at* conservation. Countries differ in their interpretation, which sometimes makes comparisons difficult: some of the sites that "count" as a protected area in one country will not necessarily be regarded as such in another. IUCN has tried to seek some measure of consensus on this issue amongst key stakeholders. While we recognise that it is up to individual countries to determine what they describe as a protected area, the weight of opinion amongst IUCN members and others seems to be towards tightening the definition overall.

One implication is that not all areas that are valuable to conservation – for instance well managed forests, sustainable use areas, military training areas or various forms of broad landscape designation – will be "protected areas" as recognised by IUCN. It is not our intention to belittle or undermine such wider efforts at sustainable management. We recognise that these management approaches are valuable for conservation, but they fall outside IUCN's definition of a protected area as set out in these guidelines.

1. Background

The first section of the guidelines sets the scene by introducing what IUCN means by the term "protected area". It looks at the history of the IUCN protected area categories, including the current process of revising the guidelines. It then explains the main purposes of the categories as understood by IUCN. Finally, a glossary gives definitions of key terms that are used in the guidelines to ensure consistency in understanding.

Protected areas

Protected areas are essential for biodiversity conservation. They are the cornerstones of virtually all national and international conservation strategies, set aside to maintain functioning natural ecosystems, to act as refuges for species and to maintain ecological processes that cannot survive in most intensely managed landscapes and seascapes. Protected areas act as benchmarks against which we understand human interactions with the natural world. Today they are often the only hope we have of stopping many threatened or endemic species from becoming extinct. They are complementary to measures to achieve conservation and sustainable use of biodiversity outside protected areas in accordance with CBD guidelines such as the Malawi and Addis Ababa Principles (CBD VII/11–12). Most protected areas exist in natural or near-natural ecosystems, or are being restored to such a state, although there are exceptions. Many contain major features of earth history and earth processes while others document the subtle interplay between human activity and nature in cultural landscapes. Larger and more natural protected areas also provide space for evolution and future ecological adaptation and restoration, both increasingly important under conditions of rapid climate change.

Such places also have direct human benefits. People – both those living in or near protected areas and others from further away – gain from the opportunities for recreation and renewal available in national parks and wilderness areas, from the genetic potential of wild species, and the environmental services provided by natural ecosystems, such as provision of water. Many protected areas are also essential for vulnerable human societies and conserve places of value such as sacred natural sites. Although many protected areas are set up by governments, others are increasingly established by local communities, indigenous peoples, environmental charities, private individuals, companies and others.

There is a huge and growing interest in the natural world, and protected areas provide us with opportunities to interact with nature in a way that is increasingly difficult elsewhere. They give us space that is otherwise lacking in an increasingly managed and crowded planet.

Protected areas also represent a commitment to future generations. Most people also believe that we have an ethical obligation to prevent species loss due to our own actions and this is supported by the teachings of the large majority of the world's religious faiths (Dudley *et al.* 2006). Protecting iconic landscapes and seascapes is seen as being important from a wider cultural perspective as well, and flagship protected areas are as important to a country's heritage as, for example, famous buildings such as the Notre Dame Cathedral or the Taj Mahal, or national football teams or works of art.

Growth in the world's protected areas system

Today roughly a tenth of the world's land surface is under some form of protected area. Over the last 40 years the global protected area estate has increased from an area the size of the United Kingdom to an area the size of South America. However, significant challenges remain. Many protected areas are not yet fully implemented or managed. Marine protected areas are lagging far behind land and inland water protected areas although there are now great efforts to rectify this situation. The vast majority of protected areas were identified and gazetted during the twentieth century, in what is almost certainly the largest and fastest conscious change of land management in history (although not as large as the mainly unplanned land degradation that has taken place over the same period). This shift in values has still to be fully recognised and understood. Protected areas continue to be established, and received a boost in 2004 when the Convention on Biological Diversity (CBD) agreed an ambitious *Programme of Work on Protected Areas*, based on the key outcomes from the Vth IUCN World Parks Congress,[1] which aims to complete ecologically-representative protected area systems around the world and has almost a hundred time-limited targets. This is necessary because although the rate of growth has been impressive, many protected areas have been set up in remote, unpopulated or only sparsely populated areas such as mountains, ice-fields and tundra and there are still notable gaps in protected area systems in some forest and grassland ecosystems, in deserts and semi-deserts, in fresh waters and, particularly, in coastal and marine areas. Many of the world's wild plant and animal species do not have viable populations in protected areas and a substantial proportion remain completely outside protected areas (Rodrigues *et al.* 2004). New protected areas are therefore likely to continue to be established in the future. One important development in the last decade is the increasing professionalism of protected area selection, through use of techniques such as ecological gap analysis (Dudley and Parrish 2006).

At the same time, there has been a rapid increase in our understanding of how such areas should be managed. In the rush to establish protected areas, often to save fragments of natural land and water from a sudden onslaught of development, protected areas were often set aside without careful analysis of the skills and capacity needed to maintain them. Knowledge is growing fast at all levels of management, from senior planners to field rangers, and there is an increasingly sophisticated volunteer network prepared to support the development of protected area systems. In a parallel development, many local communities and traditional

---

[1] Held in Durban, South Africa in September 2003.

and indigenous peoples are starting to see protected areas as one way of protecting places that are important to them, for instance sacred natural sites or areas managed for environmental benefits such as clean water or maintenance of fish stocks.

The variety of protection

The term "protected area" is therefore shorthand for a some-times bewildering array of land and water designations, of which some of the best known are *national park*, *nature reserve*, *wilderness area*, *wildlife management area* and *landscape protected area* but can also include such approaches as *community conserved areas*. More importantly, the term embraces a wide range of different management approaches, from highly protected sites where few if any people are allowed to enter, through parks where the emphasis is on conservation but visitors are welcome, to much less restrictive approaches where conservation is integrated into the traditional (and sometimes not so traditional) human lifestyles or even takes place alongside limited sustainable resource extraction. Some protected areas ban activities like food collecting, hunting or extraction of natural resources while for others it is an accepted and even a necessary part of management. The approaches taken in terrestrial, inland water and marine protected areas may also differ significantly and these differences are spelled out later in the guidelines.

The variety reflects recognition that conservation is not achieved by the same route in every situation and what may be desirable or feasible in one place could be counter-productive or politically impossible in another. Protected areas are the result of a welcome emphasis on long-term thinking and care for the natural world but also sometimes come with a price tag for those living in or near the areas being protected, in terms of lost rights, land or access to resources. There is increasing and very justifiable pressure to take proper account of human needs when setting up protected areas and these sometimes have to be "traded off" against conservation needs. Whereas in the past, governments often made decisions about protected areas and informed local people afterwards, today the emphasis is shifting towards greater discussions with stakeholders and joint decisions about how such lands should be set aside and managed. Such negotiations are never easy but usually produce stronger and longer-lasting results for both conservation and people.

IUCN recognises that many approaches to establishing and managing protected areas are valid and can make substantive contributions to conservation strategies. This does not mean that they are all equally useful in every situation: skill in selecting and combining different management approaches within and between protected areas is often the key to developing an effective functioning protected area system. Some situations will need strict protection; others can function with, or do better with, less restrictive management approaches or zoning of different management strategies within a single protected area.

Describing different approaches

In an attempt to make sense of and to describe the different approaches, IUCN has agreed a ***definition*** of what a protected area is and is not, and then identified six different protected area ***categories***, based on management objectives, one of which is subdivided into two parts. Although the categories were originally intended mainly for the reasonably modest aim of helping to collate data and information on protected areas, they have grown over time into a more complex tool. Today the categories both encapsulate IUCN's philosophy of protected areas and also help to provide a framework in which various protection strategies can be combined together, along with supportive management systems outside protected areas, into a coherent approach to conserving nature. The IUCN categories are now used for purposes as diverse as planning, setting regulations, and negotiating land and water uses. This book describes the categories and explains how they can be used to plan, implement and assess conservation strategies.

A word of warning: protected areas exist in an astonishing variety – in size, location, management approaches and objectives. Any attempt to squash such a rich and complicated collection into half a dozen neat little boxes can only ever be approximate. The IUCN protected area definition and categories are not a straitjacket but a framework to guide improved application of the categories.

History of the IUCN protected area categories

As protected areas in the modern sense were set up in one country after another during the twentieth century, each nation developed its own approach to their management and there were initially no common standards or terminology. One result is that many different terms are used at the national level to describe protected areas and there are also a variety of international protected area systems created under global conventions (e.g., World Heritage sites) and regional agreements (e.g., Natura 2000 sites in Europe).

The first effort to clarify terminology was made in 1933, at the International Conference for the Protection of Fauna and Flora, in London. This set out four protected area categories: *national park*; *strict nature reserve*; *fauna and flora reserve*; and *reserve with prohibition for hunting and collecting*. In 1942, the Western Hemisphere Convention on Nature Protection and Wildlife Preservation also incorporated four types: *national park*; *national reserve*; *nature monument*; and *strict wilderness reserve* (Holdgate 1999).

In 1962, IUCN's newly formed Commission on National Parks and Protected Areas (CNPPA), now the World Commission on Protected Areas (WCPA), prepared a *World List of*

National Parks and Equivalent Reserves, for the First World Conference on National Parks in Seattle, with a paper on nomenclature by C. Frank Brockman (1962). In 1966, IUCN produced a second version of what became a regular publication now known as the *UN List of Protected Areas*, using a simple classification system: *national parks*, *scientific reserves* and *natural monuments*. The 1972 Second World Parks Conference called on IUCN to "*define the various purposes for which protected areas are set aside; and develop suitable standards and nomenclature for such areas*" (Elliott 1974).

This was the background to the CNPPA decision to develop a categories system for protected areas. A working group report (IUCN 1978) argued that a categorization system should: show how national parks can be complemented by other types of protected area; help nations to develop management categories to reflect their needs; help IUCN to assemble and analyse data on protected areas; remove ambiguities and inconsistencies; and ensure that "*regardless of nomenclature used by nations … a conservation area can be recognised and categorised by the objectives for which it is in fact managed*". Ten categories were proposed, defined mainly by management objective, all of which were considered important, with no category inherently more valuable than another:

Group A: Categories for which CNPPA will take special responsibility
I   Scientific reserve
II   National park
III   Natural monument/national landmark
IV   Nature conservation reserve
V   Protected landscape

Group B: Other categories of importance to IUCN, but not exclusively in the scope of CNPPA
VI   Resource reserve
VII   Anthropological reserve
VIII Multiple-use management area

Group C: Categories that are part of international programmes
IX   Biosphere reserve
X   World Heritage site (natural)

However, limitations in the system soon became apparent. It did not contain a definition of a protected area; several terms were used to describe the entire suite of ten categories; a single protected area could be in more than one category; and the system lacked a marine dimension.

Revision and proposals for new categories

In 1984 CNPPA established a task force to update the categories. This reported in 1990, advising that a new system be built around the 1978 categories I–V, whilst abandoning categories

VI–X (Eidsvik 1990). CNPPA referred this to the 1992 World Parks Congress in Caracas, Venezuela. A three-day workshop there proposed maintaining a category that would be close to what had previously been category VIII for protected areas where sustainable use of natural resources was an objective. The Congress supported this and in January 1994, the IUCN General Assembly meeting in Buenos Aires approved the new system. Guidelines were published by IUCN and the World Conservation Monitoring Centre later that year (IUCN 1994). These set out a definition of a "protected area" – *An area of land and/or sea especially dedicated to the protection and maintenance of biological diversity, and of natural and associated cultural resources, and managed through legal or other effective means* – and six categories:

Areas managed mainly for:
I   Strict protection [Ia) Strict nature reserve and Ib) Wilderness area]
II   Ecosystem conservation and protection (i.e., National park)
III   Conservation of natural features (i.e., Natural monument)
IV   Conservation through active management (i.e., Habitat/species management area)
V   Landscape/seascape conservation and recreation (i.e., Protected landscape/seascape)
VI   Sustainable use of natural resources (i.e., Managed resource protected area)

The 1994 guidelines are based on key principles: the basis of categorization is by primary management objective; assignment to a category is not a commentary on management effectiveness; the categories system is international; national names for protected areas may vary; all categories are important; and a gradation of human intervention is implied.

Developments since 1994

Since publication of the guidelines, IUCN has actively promoted the understanding and use of the categories system. It has been involved in publications on how to apply the guidelines in specific geographical or other contexts (e.g., EUROPARC and IUCN 1999; Bridgewater *et al.* 1996) and a specific volume of guidelines for category V protected areas (Phillips 2002). The categories system was the cornerstone of a WCPA position statement on mining and protected areas, which was taken up in a recommendation (number 2.82) adopted by the IUCN World Conservation Congress in Amman in 2000.

IUCN secured the endorsement of the system by the Convention on Biological Diversity, at the 7th Conference of the Parties to the CBD in Kuala Lumpur in February 2004. At the Durban Worlds Parks Congress (2003) and the Bangkok World Conservation Congress (2004), proposals were made to add a governance dimension to the categories.

Finally, IUCN supported a research project by Cardiff University, UK on the use and performance of the 1994 system: *Speaking a Common Language*. The results were discussed in draft at the 2003 World Parks Congress and published for the 2004 World Conservation Congress (Bishop *et al.* 2004). A digest of papers was also published in *PARKS* in 2004 (IUCN 2004). This project helped to bring the WCPA Categories Task Force into being and to initiate the review process that has resulted in the new set of guidelines.

The current process of revision

The current guidelines are the result of an intensive process of consultation and revision coordinated by a specially appointed task force of WCPA, working closely with WCPA members and also with the other five IUCN commissions. The task force drew up its initial work plan from the results of the *Speaking a Common Language* project but with a wider mandate from IUCN to look at all aspects of the categories. It spent 18 months collecting information, talking and listening through a series of steps:

- **Research**: many people inside and outside the WCPA network contributed to the guidelines revision by writing a series of working papers, looking at different aspects of the categories. Around 40 papers were written, ranging from discussion and challenge papers through to papers that made very specific proposals or suggested text for the new guidelines. Together they form an important resource that looks at the way in which a range of protected area management objectives contribute to conservation.
- **Meetings and discussion**: the task force carried out a series of meetings around the world, or contributed to existing meetings, to give people the chance to talk about their opinions, hopes and concerns about approaches to managing protected areas. Key meetings included:
  - **Category V**: joint meeting with the WCPA Landscapes Task Force in Catalonia, Spain in 2006, supported by the Catalonian government to develop a position on category V and landscape approaches, followed by a further meeting of the Task Force in North Yorkshire, England in 2008;
  - **Category VI**: meeting in Brazil to prepare a position paper and plan a technical manual in 2007;
  - **Europe**: discussion at the European WCPA meeting in Barcelona to draw together opinions from European WCPA members in 2007;
  - **South and East Africa**: two-day workshop in Nairobi in 2006 in collaboration with UNEP-WCMC, attended by representative from 13 African states;
  - **South-East Asia**: two-day workshop on governance and categories at a regional conference in Kota Kinabalu in Sabah, Malaysia in 2007 with representatives from 17 countries;
  - **Latin America**: discussions at the Latin American protected areas congress at Bariloche, Argentina in

2007, focusing in particular on issues relating to category VI, marine protected areas and indigenous reserves;
  - **International Council on Metals and Mining**: presentation followed by discussion leading to a working paper from ICMM members during 2007.
  - There were also a series of smaller meetings: e.g., with the IUCN UK Committee, Canadian Council for Ecological Areas, WWF Conservation Science Programme, Conservation International, UNESCO, industry stakeholders at IUCN headquarters etc.
  - In addition, there was a **global "summit"** on protected area categories in Spain in May 2007, funded and supported technically by the Andalusia regional government, the Spanish Ministry of the Environment and "Fundación Biodiversidad". It was attended by over a hundred experts from around the world, with four days to discuss a wide range of issues relating to the categories. Although this was not a decision-making meeting, the various consensus positions developed during the meeting helped to set the form of the revised guidelines.
- **Website**: The task force has a dedicated site on the WCPA website, with all relevant papers etc. available: www.iucn.org/themes/wcpa/theme/categories/about.html
- **E-forum**: In the run-up to the summit, IUCN and the task force coordinated a E-discussion open to everyone about the categories, which provided invaluable input to the thinking about the next stages in the revision process.

Draft guidelines were prepared for the Steering Committee meeting of the World Commission on Protected Areas in September 2007, and revised following comments from Steering Committee members. The various drafts were produced in English only, a limitation created by shortage of funds, although the final guidelines are being published in full in English, French and Spanish, with summaries in other languages. Guidelines were made available to all WCPA members and any other interested parties for comment, and many comments were received and incorporated into the text. A separate consultation was made related to the protected area definition.

The WCPA Steering Committee met again in April 2008 in Cape Town and discussed the draft in detail both in open session and in break-out groups to address particular issues. Final decisions about what to propose to IUCN membership were made where necessary by the chair of WCPA.

Purpose of the IUCN protected area management categories

IUCN sees the protected area management categories as an important global standard for the planning, establishment and management of protected areas; this section outlines the main uses recognised. These have developed since the original category guidelines were published in 1994 and the list of possible uses is longer. On the other hand, the categories are sometimes used as tools beyond their original aims, perhaps in the absence of any alternative, and we need to distinguish uses that IUCN supports and those that it is neutral about or opposed to.

Purposes that IUCN supports and actively encourages

Facilitating planning of protected areas and protected area systems

- To provide a tool for planning protected area systems and wider bioregional or ecoregional conservation planning exercises;
- To encourage governments and other owners or managers of protected areas to develop systems of protected areas with a range of management objectives tailored to national and local circumstances;
- To give recognition to different management arrangements and governance types.

Improving information management about protected areas

- To provide international standards to help global and regional data collection and reporting on conservation efforts, to facilitate comparisons between countries and to set a framework for global and regional assessments;
- To provide a framework for the collection, handling and dissemination of data about protected areas;
- To improve communication and understanding between all those engaged in conservation;
- To reduce the confusion that has arisen from the adoption of many different terms to describe the same kinds of protected areas in different parts of the world.

Helping to regulate activities in protected areas

- To use the categories as guidelines on a national or international level to help regulate activities e.g., by prescribing certain activities in some categories in accordance with the management objectives of the protected area.

Purposes that are becoming increasingly common, that IUCN supports and on which it is prepared to give advice

- To provide the basis for legislation – a growing number of countries are using the IUCN categories as a or the basis for categorizing protected areas under law;
- To set budgets – some countries base scales of annual budgets for protected areas on their category;
- To use the categories as a tool for advocacy – NGOs are using categories as a campaign tool to promote conservation objectives and appropriate levels of human use activities;
- To interpret or clarify land tenure and governance – some indigenous and local communities are using the categories as a tool to help to establish management systems such as indigenous reserves;
- To provide tools to help plan systems of protected areas with a range of management objectives and governance types.

Purposes that IUCN opposes

- To use the categories as an excuse for expelling people from their traditional lands;
- To change categories to downgrade protection of the environment;
- To use the categories to argue for environmentally insensitive development in protected areas.

2. Definition and categories

This section outlines and explains the IUCN definition of a protected area, a protected area system and the six categories. The definition is clarified phrase by phrase and should be applied with some accompanying principles. Categories are described by their main objective, other objectives, distinguishing features, role in the landscape or seascape, unique points and actions that are compatible or incompatible.

The new IUCN definition of a protected area

> The IUCN definition is given and explained, phrase by phrase

IUCN members have worked together to produce a revised definition of a protected area, which is given below. The first draft of this new definition was prepared at a meeting on the categories in Almeria, Spain in May 2007 and since then has been successively refined and revised by many people within IUCN-WCPA.

> A protected area is: **"A clearly defined geographical space, recognised, dedicated and managed, through legal or other effective means, to achieve the long-term conservation of nature with associated ecosystem services and cultural values"**.
>
> In applying the categories system, the first step is to determine whether or not the site meets this definition and the second step is to decide on the most suitable category.

This definition packs a lot into one short sentence. Table 1 looks at each word and/or phrase in turn and expands on the meaning.

Table 1.   **Explanation of protected area definition**

Phrase	Explanation	Examples and further details
Clearly defined geographical space	Includes land, inland water, marine and coastal areas or a combination of two or more of these. "Space" has three dimensions, e.g., as when the airspace above a protected area is protected from low-flying aircraft or in marine protected areas when a certain water depth is protected or the seabed is protected but water above is not: conversely subsurface areas sometimes are *not* protected (e.g., are open for mining). "Clearly defined" implies a spatially defined area with agreed and demarcated borders. These borders can sometimes be defined by physical features that move over time (e.g., river banks) or by management actions (e.g., agreed no-take zones).	**Wolong Nature Reserve** in China (category Ia, terrestrial); **Lake Malawi National Park** in Malawi (category II, mainly freshwater); **Masinloc and Oyon Bay Marine Reserve** in the Philippines (category Ia, mainly marine) are examples of areas in very different biomes but all are protected areas.
Recognised	Implies that protection can include a range of governance types declared by people as well as those identified by the state, but that such sites should be recognised in some way (in particular through listing on the World Database on Protected Areas – WDPA).	**Anindilyakwa Indigenous Protected Area** (IPA) was self-declared by aboriginal communities in the Groote Eylandt peninsula, one of many self-declared IPAs recognised by the government.
Dedicated	Implies specific binding commitment to conservation in the long term, through e.g.: • International conventions and agreements • National, provincial and local law • Customary law • Covenants of NGOs • Private trusts and company policies • Certification schemes.	Cradle Mountain – **Lake St Clair National Park** in Tasmania, Australia (category II, state); **Nabanka Fish Sanctuary** in the Philippines (community conserved area); **Port Susan Bay Preserve** in Washington, USA (private) are all protected areas, but their legal structure differs considerably.
Managed	Assumes some active steps to conserve the natural (and possibly other) values for which the protected area was established; note that "managed" can include a decision to leave the area untouched if this is the best conservation strategy.	Many options are possible. For instance **Kaziranga National Park** in India (category II) is managed mainly through poaching controls and removal of invasive species; islands in the **Archipelago National Park** in Finland are managed using traditional farming methods to maintain species associated with meadows.
Legal or other effective means	Means that protected areas must either be gazetted (that is, recognised under statutory civil law), recognised through an international convention or agreement, or else managed through other effective but non-gazetted means, such as through recognised traditional rules under which community conserved areas operate or the policies of established non-governmental organizations.	**Flinders Range National Park** in Australia is managed by the state authority of South Australia; **Attenborough Nature Reserve** in the UK is managed by the county Nottinghamshire Wildlife Trust in association with the gravel company that owns the site; and the **Alto Fragua Indiwasi National Park** in Colombia is managed by the Ingano peoples.

Table 1. **Explanation of protected area definition (cont.)**

Phrase	Explanation	Examples and further details
… to achieve	Implies some level of effectiveness – a new element that was not present in the 1994 definition but which has been strongly requested by many protected area managers and others. Although the category will still be determined by objective, management effectiveness will progressively be recorded on the World Database on Protected Areas and over time will become an important contributory criterion in identification and recognition of protected areas.	The **Convention on Biological Diversity** is asking Parties to carry out management effectiveness assessments.
Long-term	Protected areas should be managed in perpetuity and not as a short-term or temporary management strategy.	Temporary measures, such as short-term grant-funded agricultural set-asides, rotations in commercial forest management or temporary fishing protection zones are not protected areas as recognised by IUCN.
Conservation	In the context of this definition conservation refers to the *in-situ* maintenance of ecosystems and natural and semi-natural habitats and of viable populations of species in their natural surroundings and, in the case of domesticated or cultivated species (see definition of agrobiodiversity in the Appendix), in the surroundings where they have developed their distinctive properties.	**Yellowstone National Park** in the United States (category II) has conservation aims focused in particular on maintaining viable populations of bears and wolves but with wider aims of preserving the entire functioning ecosystem.
Nature	In this context nature *always* refers to biodiversity, at genetic, species and ecosystem level, and often *also* refers to geodiversity, landform and broader natural values.	**Bwindi Impenetrable Forest National Park** in Uganda (category II) is managed primarily to protect natural mountain forests and particularly the mountain gorilla. The **Island of Rum National Nature Reserve** in Scotland (category IV) was set up to protect unique geological features.
Associated ecosystem services	Means here ecosystem services that are related to but do not interfere with the aim of nature conservation. These can include provisioning services such as food and water; regulating services such as regulation of floods, drought, land degradation, and disease; supporting services such as soil formation and nutrient cycling; and cultural services such as recreational, spiritual, religious and other non-material benefits.	Many protected areas also supply ecosystem services: e.g., **Gunung Gede National Park** in Java, Indonesia (category II) helps supply fresh water to Jakarta; and the **Sundarbans National Park** in Bangladesh (category IV) helps to protect the coast against flooding.
Cultural values	Includes those that do not interfere with the conservation outcome (*all* cultural values in a protected area should meet this criterion), including in particular: • those that contribute to conservation outcomes (e.g., traditional management practices on which key species have become reliant); • those that are themselves under threat.	Many protected areas contain sacred sites, e.g., **Nyika National Park** in Malawi has a sacred pool, waterfall and mountain. Traditional management of forests to supply timber for temples in Japan has resulted in some of the most ancient forests in the country, such as the protected primeval forest outside **Nara**. The **Kaya** forests of coastal Kenya are protected both for their biodiversity and their cultural values.

The three-dimensional aspects of protected areas

In some situations protected areas need to consider the impacts of human activities in three dimensions. Issues can include: protecting the airspace above a protected area for instance from disturbance from low-flying aircraft, helicopter flights or hot-air balloons; and limiting human activity below the surface such as mining and other extractive industries. Issues specific to marine and inland water sites include fishing, dredging, diving

and underwater noise. A number of countries have enshrined three-dimensional aspects into their protected area legislation; for example Cuba bans mining below protected areas. IUCN encourages governments to consider a general legal provision to safeguard protected areas from intrusive activities above and/or below ground and underwater. It encourages governments to ensure that assessments are undertaken to ascertain the potential effects of such activities before any decisions are taken on whether they should be permitted and if so whether particular limits or conditions should apply.

Principles

> The definition should be applied in the context of a series of accompanying principles, outlined below

- For IUCN, only those areas where the main objective is conserving nature can be considered protected areas; this can include many areas with other goals as well, at the same level, but in the case of conflict, nature conservation will be the priority;
- Protected areas must prevent, or eliminate where necessary, any exploitation or management practice that will be harmful to the objectives of designation;
- The choice of category should be based on the primary objective(s) stated for each protected area;
- The system is not intended to be hierarchical;
- All categories make a contribution to conservation but objectives must be chosen with respect to the particular situation; not all categories are equally useful in every situation;
- Any category can exist under any governance type and *vice versa*;
- A diversity of management approaches is desirable and should be encouraged, as it reflects the many ways in which communities around the world have expressed the universal value of the protected area concept;
- The category should be changed if assessment shows that the stated, long-term management objectives do not match those of the category assigned;
- However, the category is not a reflection of management effectiveness;
- Protected areas should usually aim to maintain or, ideally, increase the degree of naturalness of the ecosystem being protected;
- The definition and categories of protected areas should not be used as an excuse for dispossessing people of their land.

Definition of a protected area system and the ecosystem approach

> The categories should be applied in the context of national or other protected area systems and as part of the ecosystem approach

IUCN emphasises that protected areas should not be seen as isolated entities, but part of broader conservation landscapes, including both protected area systems and wider ecosystem approaches to conservation that are implemented across the landscape or seascape. The following section provides outline definitions of both these concepts.

Protected area system

The overriding purpose of a system of protected areas is to increase the effectiveness of *in-situ* biodiversity conservation.

IUCN has suggested that the long-term success of *in-situ* conservation requires that the global system of protected areas comprise a representative sample of each of the world's different ecosystems (Davey 1998). IUCN WCPA characterizes a protected area system as having five linked elements (Davey 1998 with additions):

- **Representativeness, comprehensiveness and balance**: including highest quality examples of the full range of environment types within a country; includes the extent to which protected areas provide balanced sampling of the environment types they purport to represent.
- **Adequacy**: integrity, sufficiency of spatial extent and arrangement of contributing units, together with effective management, to support viability of the environmental processes and/or species, populations and communities that make up the biodiversity of the country.
- **Coherence and complementarity**: positive contribution of each protected area towards the whole set of conservation and sustainable development objectives defined for the country.
- **Consistency**: application of management objectives, policies and classifications under comparable conditions in standard ways, so that the purpose of each protected area within the system is clear to all and to maximize the chance that management and use support the objectives.
- **Cost effectiveness, efficiency and equity**: appropriate balance between the costs and benefits, and appropriate equity in their distribution; includes efficiency: the minimum number and area of protected areas needed to achieve system objectives.

In 2004, the CBD *Programme of Work on Protected Areas* provided some criteria for protected area systems in the Programme's overall objective to establish and maintain "*comprehensive, effectively managed, and ecologically representative national and regional systems of protected areas*".

Ecosystem approaches

IUCN believes that protected areas should be integrated into coherent protected area systems, and that such systems should further be integrated within broader-scale approaches to conservation and land/water use, which include both protected land and water and a wide variety of sustainable management approaches. This is in line with the CBD Malawi Principles (CBD/COP4, 1998) noting the importance of sustainable use strategies. These broader-scale conservation strategies are called variously "landscape-scale approaches", "bioregional approaches" or "ecosystem approaches". Where such approaches include the conservation of areas that connect protected areas the term "connectivity conservation" is used. Individual protected areas should therefore wherever possible contribute to national and regional protected areas and broad-scale conservation plans.

The ecosystem approach is a broader framework for planning and developing conservation and land/water use management in an integrated manner. In this context, protected areas fit as one important tool – perhaps the most important tool – in such an approach.

The CBD defines the ecosystem approach as: "*a strategy for the integrated management of land, water and living resources that promotes conservation and sustainable use in an equitable way …* " (CBD 2004).

Categories

The individual categories are described in turn under a series of headings:

- Primary objective(s)
- Other objectives
- Distinguishing features
- Role in the landscape or seascape
- What makes the category unique
- Issues for consideration

Names of protected areas

The categories system was introduced in large part to help standardize descriptions of what constitutes a particular protected area. **The names of all protected areas except the ones in category II were chosen to relate, more or less closely, to the main management objective of the category**.

The term "National Park", which existed long before the categories system, was found to apply particularly well to large protected areas under category II. It is true however, that many existing national parks all over the world have very different aims from those defined under category II. As a matter of fact, some countries have categorized their national parks under other IUCN categories (see Table 2 below).

Table 2.   "National parks" in various categories

Category	Name	Location	Size (ha)	Date
Ia	Dipperu National Park	Australia	11,100	1969
II	Guanacaste National Park	Costa Rica	32,512	1991
III	Yozgat Camligi National Park	Turkey	264	1988
IV	Pallas Ounastunturi National Park	Finland	49,600	1938
V	Snowdonia National Park	Wales, UK	214,200	1954
VI	Expedition National Park	Australia	2,930	1994

It is important to note that **the fact that a government has called, or wants to call, an area a national park does not mean that it has to be managed according to the guidelines under category II**. Instead the most suitable management system should be identified and applied; the name is a matter for governments and other stakeholders to decide.

What follows is a framework. Although some protected areas will fall naturally into one or another category, in other cases the distinctions will be less obvious and will require in-depth analysis of options. Because assignment of a category depends on management objective, it depends more on what the management authority *intends* for the site rather than on any strict and inviolable set of criteria. Some tools are available to help make the decision about category, but in many cases the final decision will be a matter of collective judgement.

In addition, because the system is global, it is also inevitably fairly general. IUCN encourages countries to add greater detail to definition of the categories for their own national circumstances if this would be useful, keeping within the general guidelines outlined below. Several countries have already done this or are in the process of doing so and IUCN encourages this process.

Natural and cultural landscapes/seascapes

We note that few if any areas of the land, inland waters and coastal seas remain completely unaffected by direct human activity, which has also impacted on the world's oceans through fishing pressure and pollution. If the impacts of transboundary air pollution and climate change are factored in, the entire planet has been modified. It therefore follows that terms such as "natural" and "cultural" are approximations. To some extent we could describe all protected areas as existing in "cultural" landscapes in that cultural practices will have changed and influenced ecology, often over millennia. However, this is little help in distinguishing between very different types of ecosystem functioning. We therefore use the terms as follows:

Natural or unmodified areas are those that still retain a complete or almost complete complement of species native to the area, within a more-or-less naturally functioning ecosystem.

Cultural areas have undergone more substantial changes by, for example, settled agriculture, intensive permanent grazing and forest management that have altered the composition or structure of the forest. Species composition and ecosystem functioning are likely to have been substantially altered. Cultural landscapes can however still contain a rich array of species and in some cases these may have become reliant on cultural management.

Use of terms such as "natural" and "un-modified" does not seek to hide or deny the long-term stewardship of indigenous and traditional peoples where this exists; indeed many areas remain valuable to biodiversity precisely because of this form of management.

Objectives common to all six protected area categories

The definition implies a common set of objectives for protected areas; the categories in turn define differences in management approaches. The following objectives should or can apply to all protected area categories: i.e., they do not distinguish any one category from another.

All protected areas should aim to:
- Conserve the composition, structure, function and evolutionary potential of biodiversity;

- Contribute to regional conservation strategies (as core reserves, buffer zones, corridors, stepping-stones for migratory species etc.);

- Maintain diversity of landscape or habitat and of associated species and ecosystems;

- Be of sufficient size to ensure the integrity and long-term maintenance of the specified conservation targets or be capable of being increased to achieve this end;

- Maintain the values for which it was assigned in perpetuity;

- Be operating under the guidance of a management plan, and a monitoring and evaluation programme that supports adaptive management;

- Possess a clear and equitable governance system.

All protected areas should also aim where appropriate[2] to:
- Conserve significant landscape features, geomorphology and geology;

- Provide regulatory ecosystem services, including buffering against the impacts of climate change;

- Conserve natural and scenic areas of national and international significance for cultural, spiritual and scientific purposes;

- Deliver benefits to resident and local communities consistent with the other objectives of management;

- Deliver recreational benefits consistent with the other objectives of management;

- Facilitate low-impact scientific research activities and ecological monitoring related to and consistent with the values of the protected area;

- Use adaptive management strategies to improve management effectiveness and governance quality over time;

- Help to provide educational opportunities (including about management approaches);

- Help to develop public support for protection.

It should be noted that IUCN's members adopted a recommendation at the World Conservation Congress in Amman, Jordan in October 2000, which suggested that mining should not take place in IUCN category I–IV protected areas. Recommendation 2.82 includes a section that: "*Calls on all IUCN's*

---

[2] This distinction is made because not all protected areas will contain significant geology, ecosystem services, opportunities for local livelihoods etc., so such objectives are not universal, but are appropriate whenever the opportunity occurs. The following pages describe distinct features of each management category that add to these basic aims. In some cases an objective such as scientific research or recreation may be mentioned because it is a major aim of a particular category.

State members to prohibit by law, all exploration and extraction of mineral resources in protected areas corresponding to IUCN protected area management categories I–IV". The recommendation also includes a paragraph relating to category V and VI protected areas: "*in categories V and VI, exploration and localized extraction would be accepted only where the nature and extent of the proposed activities of the mining project indicate the compatibility of the project activities with the objectives of the protected areas*". This is a recommendation and not in any way binding on governments; some currently do ban mining in categories I–IV protected areas and others do not.

Category Ia: Strict nature reserve

> **Category Ia** are strictly protected areas set aside to protect biodiversity and also possibly geological/geomorphological features, where human visitation, use and impacts are strictly controlled and limited to ensure protection of the conservation values. Such protected areas can serve as indispensable reference areas for scientific research and monitoring.

Before choosing a category, check first that the site meets the definition of a protected area (page 8).

Primary objective

- To conserve regionally, nationally or globally outstanding ecosystems, species (occurrences or aggregations) and/or geodiversity features: these attributes will have been formed mostly or entirely by non-human forces and will be degraded or destroyed when subjected to all but very light human impact.

Other objectives

- To preserve ecosystems, species and geodiversity features in a state as undisturbed by recent human activity as possible;
- To secure examples of the natural environment for scientific studies, environmental monitoring and education, including baseline areas from which all avoidable access is excluded;
- To minimize disturbance through careful planning and implementation of research and other approved activities;
- To conserve cultural and spiritual values associated with nature.

Distinguishing features

The area should generally:

- Have a largely complete set of expected native species in ecologically significant densities or be capable of returning them to such densities through natural processes or time-limited interventions;

- Have a full set of expected native ecosystems, largely intact with intact ecological processes, or processes capable of being restored with minimal management intervention;
- Be free of significant direct intervention by modern humans that would compromise the specified conservation objectives for the area, which usually implies limiting access by people and excluding settlement;
- Not require substantial and on-going intervention to achieve its conservation objectives;
- Be surrounded when feasible by land uses that contribute to the achievement of the area's specified conservation objectives;
- Be suitable as a baseline monitoring site for monitoring the relative impact of human activities;
- Be managed for relatively low visitation by humans;
- Be capable of being managed to ensure minimal disturbance (especially relevant to marine environments).

The area could be of religious or spiritual significance (such as a sacred natural site) so long as biodiversity conservation is identified as a primary objective. In this case the area might contain sites that could be visited by a limited number of people engaged in faith activities consistent with the area's management objectives.

Role in the landscape/seascape

Category Ia areas are a vital component in the toolbox of conservation. As the Earth becomes increasingly influenced by human activities, there are progressively fewer areas left where such activities are strictly limited. Without the protection accompanying the Ia designation, there would rapidly be no such areas left. As such, these areas contribute in a significant way to conservation through:

- Protecting some of the earth's richness that will not survive outside of such strictly protected settings;
- Providing reference points to allow baseline and long-term measurement and monitoring of the impact of human-induced change outside such areas (e.g., pollution);
- Providing areas where ecosystems can be studied in as pristine an environment as possible;
- Protecting additional ecosystem services;
- Protecting natural sites that are also of religious and cultural significance.

What makes category Ia unique?

Allocation of category is a matter of choice, depending on long-term management objectives, often with a number of alternative options that could be applied in any one site. The following box outlines some of the main reasons why Category Ia may be chosen in specific situations *vis-à-vis* other categories that pursue similar objectives.

Category Ia differs from the other categories in the following ways:	
Category Ib	Category Ib protected areas will generally be larger and less strictly protected from human visitation than category Ia: although not usually subject to mass tourism they may be open to limited numbers of people prepared for self-reliant travel such as on foot or by boat, which is not always the case in Ia.
Category II	Category II protected areas usually combine ecosystem protection with recreation, subject to zoning, on a scale not suitable for category I.
Category III	Category III protected areas are generally centred on a particular natural feature, so that the primary focus of management is on maintaining this feature, whereas objectives of Ia are generally aimed at a whole ecosystem and ecosystem processes.
Category IV	Category IV protected areas protect fragments of ecosystems or habitats, which often require continual management intervention to maintain. Category Ia areas on the other hand should be largely self-sustaining and their objectives preclude such management activity or the rate of visitation common in category IV. Category IV protected areas are also often established to protect particular species or habitats rather than the specific ecological aims of category Ia.
Category V	Category V protected areas are generally cultural landscapes or seascapes that have been altered by humans over hundreds or even thousands of years and that rely on continuing intervention to maintain their qualities including biodiversity. Many category V protected areas contain permanent human settlements. All the above are incompatible with category Ia.
Category VI	Category VI protected areas contain natural areas where biodiversity conservation is linked with sustainable use of natural resources, which is incompatible with category Ia. However large category VI protected areas may contain category Ia areas within their boundaries as part of management zoning.

Issues for consideration

- There are few areas of the terrestrial and marine worlds which do not bear the hallmarks of earlier human action, though in many cases the original human inhabitants are no longer present. In many cases, category Ia areas will therefore require a process of restoration. This restoration should be through natural processes or time-limited interventions: if continual intervention is required the area would be more suitable in some other category, such as IV or V.

- There are few areas not under some kind of legal or at least traditional ownership, so that finding places that exclude human activity is often problematic.
- Some human actions have a regional and global reach that is not restricted by protected area boundaries. This is most apparent with climate and air pollution, and new and emerging diseases. In an increasingly modified ecology, it may become increasingly difficult to maintain pristine areas through non-intervention.
- Many sacred natural sites are managed in ways that are analogous to Ia protected areas for spiritual and cultural reasons, and may be located *within* both category V and VI protected areas.

Category Ib: Wilderness area

> **Category Ib** protected areas are usually large unmodified or slightly modified areas, retaining their natural character and influence, without permanent or significant human habitation, which are protected and managed so as to preserve their natural condition.

Before choosing a category, check first that the site meets the definition of a protected area (page 8).

Primary objective

- To protect the long-term ecological integrity of natural areas that are undisturbed by significant human activity, free of modern infrastructure and where natural forces and processes predominate, so that current and future generations have the opportunity to experience such areas.

Other objectives

- To provide for public access at levels and of a type which will maintain the wilderness qualities of the area for present and future generations;
- To enable indigenous communities to maintain their traditional wilderness-based lifestyle and customs, living at low density and using the available resources in ways compatible with the conservation objectives;
- To protect the relevant cultural and spiritual values and non-material benefits to indigenous or non-indigenous populations, such as solitude, respect for sacred sites, respect for ancestors etc.;
- To allow for low-impact minimally invasive educational and scientific research activities, when such activities cannot be conducted outside the wilderness area.

Distinguishing features

The area should generally:

- Be free of modern infrastructure, development and industrial extractive activity, including but not limited to roads, pipelines, power lines, cellphone towers, oil and gas platforms, offshore liquefied natural gas terminals, other permanent structures, mining, hydropower development, oil and gas extraction, agriculture including intensive livestock grazing, commercial fishing, low-flying aircraft etc., preferably with highly restricted or no motorized access.

- Be characterized by a high degree of intactness: containing a large percentage of the original extent of the ecosystem, complete or near-complete native faunal and floral assemblages, retaining intact predator-prey systems, and including large mammals.
- Be of sufficient size to protect biodiversity; to maintain ecological processes and ecosystem services; to maintain ecological refugia; to buffer against the impacts of climate change; and to maintain evolutionary processes.
- Offer outstanding opportunities for solitude, enjoyed once the area has been reached, by simple, quiet and non-intrusive means of travel (i.e., non-motorized or highly regulated motorized access where strictly necessary and consistent with the biological objectives listed above).
- Be free of inappropriate or excessive human use or presence, which will decrease wilderness values and ultimately prevent an area from meeting the biological and cultural criteria listed above. However, human presence should not be the determining factor in deciding whether to establish a category Ib area. The key objectives are biological intactness and the absence of permanent infrastructure, extractive industries, agriculture, motorized use, and other indicators of modern or lasting technology.

However, in addition they can include:

- Somewhat disturbed areas that are capable of restoration to a wilderness state, and smaller areas that might be expanded or could play an important role in a larger wilderness protection strategy as part of a system of protected areas that includes wilderness, if the management objectives for those somewhat disturbed or smaller areas are otherwise consistent with the objectives set out above.

Where the biological integrity of a wilderness area is secure and the primary objective listed above is met, the management focus of the wilderness area may shift to other objectives such as protecting cultural values or recreation, but only so long as the primary objective continues to be secure.

Role in the landscape/seascape

In many ways wilderness areas play similar roles to category II national parks in protecting large, functioning ecosystems (or at least areas where many aspects of an ecosystem can flourish). Their particular roles include:

- Protecting large mainly untouched areas where ecosystem processes, including evolution, can continue unhindered by human, including development or mass tourism;
- Protecting compatible ecosystem services;

- Protecting particular species and ecological communities that require relatively large areas of undisturbed habitat;
- Providing a "pool" of such species to help populate sustainably-managed areas surrounding the protected area;
- Providing space for a limited number of visitors to experience wilderness;
- Providing opportunities for responses to climate change including biome shift.

What makes category Ib unique?

Category Ib differs from the other categories in the following ways:	
Category Ia	Category Ia protected areas are strictly protected areas, generally with only limited human visitation. They are often (but not always) relatively small, in contrast to Ib. There would usually not be human inhabitants in category Ia, but use by indigenous and local communities takes place in many Ib protected areas.
Category II	Category Ib and II protected areas are often similar in size and in their aim to protect functioning ecosystems. But whereas II usually includes (or plans to include) use by visitors, including supporting infrastructure, in Ib visitor use is more limited and confined to those with the skills and equipment to survive unaided.
Category III	Category III is aimed at protecting a specific natural feature, which is not the aim of category Ib. Category III protected areas are frequently quite small and, like category II, aimed at encouraging visitors sometimes in large numbers; Ib sites on the other hand are generally larger and discourage anything but specialist visitors.
Category IV	Category IV protected areas are usually relatively small and certainly not complete functioning ecosystems, most will need regular management interventions to maintain their associated biodiversity: all these attributes are the reverse of conditions in Ib.
Category V	Category V protected areas comprise cultural landscapes and seascapes, shaped by (usually long-term) human intervention and usually containing sizable settled human communities. Category Ib should be in as natural a state as possible and would only contain cultural landscapes if the intention were to restore these back to near-natural conditions.
Category VI	Category VI is predicated on setting internal zoning and management regimes to support sustainable use; although wilderness areas sometimes include limited traditional use by indigenous people this is incidental to management aims rather than an intrinsic part of those aims.

Issues for consideration

- Some wilderness areas include livestock grazing by nomadic peoples and distinctions may need to be made between intensive and non-intensive grazing; however this will pose challenges if people want to increase stocking density.

Category II: National park

> **Category II** protected areas are large natural or near natural areas set aside to protect large-scale ecological processes, along with the complement of species and ecosystems characteristic of the area, which also provide a foundation for environmentally and culturally compatible spiritual, scientific, educational, recreational and visitor opportunities.

Before choosing a category, check first that the site meets the definition of a protected area (page 8).

Primary objective

- To protect natural biodiversity along with its underlying ecological structure and supporting environmental processes, and to promote education and recreation.[3]

Other objectives:

- To manage the area in order to perpetuate, in as natural a state as possible, representative examples of physiographic regions, biotic communities, genetic resources and unimpaired natural processes;
- To maintain viable and ecologically functional populations and assemblages of native species at densities sufficient to conserve ecosystem integrity and resilience in the long term;
- To contribute in particular to conservation of wide-ranging species, regional ecological processes and migration routes;
- To manage visitor use for inspirational, educational, cultural and recreational purposes at a level which will not cause significant biological or ecological degradation to the natural resources;
- To take into account the needs of indigenous people and local communities, including subsistence resource use, in so far as these will not adversely affect the primary management objective;
- To contribute to local economies through tourism.

Distinguishing features

Category II areas are typically large and conserve a functioning "ecosystem", although to be able to achieve this, the protected area may need to be complemented by sympathetic management in surrounding areas.

- The area should contain representative examples of major natural regions, and biological and environmental features or scenery, where native plant and animal species, habitats and geodiversity sites are of special spiritual, scientific, educational, recreational or tourist significance.
- The area should be of sufficient size and ecological quality so as to maintain ecological functions and processes that will allow the native species and communities to persist for the long term with minimal management intervention.
- The composition, structure and function of biodiversity should be to a great degree in a "natural" state or have the potential to be restored to such a state, with relatively low risk of successful invasions by non-native species.

Role in the landscape/seascape

Category II provides large-scale conservation opportunities where natural ecological processes can continue in perpetuity, allowing space for continuing evolution. They are often key stepping-stones for designing and developing large-scale biological corridors or other connectivity conservation initiatives required for those species (wide-ranging and/or migratory) that cannot be conserved entirely within a single protected area. Their key roles are therefore:

- Protecting larger-scale ecological processes that will be missed by smaller protected areas or in cultural landscapes;
- Protecting compatible ecosystem services;
- Protecting particular species and communities that require relatively large areas of undisturbed habitat;
- Providing a "pool" of such species to help populate sustainably-managed areas surrounding the protected area;
- To be integrated with surrounding land or water uses to contribute to large-scale conservation plans;
- To inform and excite visitors about the need for and potential of conservation programmes;
- To support compatible economic development, mostly through recreation and tourism, that can contribute to local and national economies and in particular to local communities.

Category II areas should be more strictly protected where ecological functions and native species composition are relatively intact; surrounding landscapes can have varying degrees of consumptive or non-consumptive uses but should ideally serve as buffers to the protected area.

---

[3] Note that the name "national park" is not exclusively linked to Category II. Places called national parks exist in all the categories (and there are even some national parks that are not protected areas at all). The name is used here because it is descriptive of Category II protected areas in many countries. The fact that an area is called a national park is independent of its management approach. In particular, the term "national park" should never be used as a way of dispossessing people of their land.

What makes category II unique?

Category II differs from the other categories in the following ways:	
Category Ia	Category II will generally not be as strictly conserved as category Ia and may include tourist infrastructure and visitation. However, category II protected areas will often have core zones where visitor numbers are strictly controlled, which may more closely resemble category Ia.
Category Ib	Visitation in category II will probably be quite different from in wilderness areas, with more attendant infrastructure (trails, roads, lodges etc.) and therefore probably a greater number of visitors. Category II protected areas will often have core zones where numbers of visitors are strictly controlled, which may more closely resemble category Ib.
Category III	Management in category III is focused around a single natural feature, whereas in category II it is focused on maintaining a whole ecosystem.
Category IV	Category II is aimed at maintaining ecological integrity at ecosystem scale, whereas category IV is aimed at protecting habitats and individual species. In practice, category IV protected areas will seldom be large enough to protect an entire ecosystem and the distinction between categories II and IV is therefore to some extent a matter of degree: category IV sites are likely to be quite small (individual marshes, fragments of woodland, although there are exceptions), while category II are likely to be much larger and at least fairly self-sustaining.
Category V	Category II protected areas are essentially natural systems or in the process of being restored to natural systems while category V are cultural landscapes and aim to be retained in this state.
Category VI	Category II will not generally have resource use permitted except for subsistence or minor recreational purposes.

Issues for consideration

- Concepts of naturalness are developing fast and some areas that may previously have been regarded as natural are now increasingly seen as to some extent cultural landscapes – e.g., savannah landscapes where fire has been used to maintain vegetation mosaics and thus populations of animals for hunting. The boundaries between what is regarded and managed as category II and category V may therefore change over time.
- Commercialization of land and water in category II is creating challenges in many parts of the world, in part because of a political perception of resources being "locked up" in national parks, with increasing pressure for greater recreational uses and lack of compliance by tour operators, development of aquaculture and mariculture schemes, and trends towards privatization of such areas.
- Issues of settled populations in proposed category II protected areas, questions of displacement, compensation (including for fishing communities displaced from marine and coastal protected areas), alternative livelihood options and changed approaches to management are all emerging themes.

Category III: Natural monument or feature

> **Category III** protected areas are set aside to protect a specific natural monument, which can be a landform, sea mount, submarine cavern, geological feature such as a cave or even a living feature such as an ancient grove. They are generally quite small protected areas and often have high visitor value.

Before choosing a category, check first that the site meets the definition of a protected area (page 8).

Primary objective

- To protect specific outstanding natural features and their associated biodiversity and habitats.

Other objectives

- To provide biodiversity protection in landscapes or seascapes that have otherwise undergone major changes;[4]
- To protect specific natural sites with spiritual and/or cultural values where these also have biodiversity values;
- To conserve traditional spiritual and cultural values of the site.

Distinguishing features

Category III protected areas are usually relatively small sites that focus on one or more prominent natural features and the associated ecology, rather than on a broader ecosystem. They are managed in much the same way as category II. The term "natural" as used here can refer to both wholly natural features (the commonest use) but also sometimes features that have been influenced by humans. In the latter case these sites should also always have important associated biodiversity attributes, which

---

[4] Noting that protection of specific cultural sites can often provide havens of natural or semi-natural habitat in areas that have otherwise undergone substantial modification – e.g., ancient trees around temples.

should be reflected as a priority in their management objectives if they are to be classified as a protected area rather than an historical or spiritual site. Category III protected areas could include:

- **Natural geological and geomorphological features:** such as waterfalls, cliffs, craters, caves, fossil beds, sand dunes, rock forms, valleys and marine features such as sea mounts or coral formations;
- **Culturally-influenced natural features:** such as cave dwellings and ancient tracks;
- **Natural-cultural sites:** such as the many forms of sacred natural sites (sacred groves, springs, waterfalls, mountains, sea coves etc.) of importance to one or more faith groups;
- **Cultural sites with associated ecology:** where protection of a cultural site also protects significant and important biodiversity, such as archaeological/historical sites that are inextricably linked to a natural area.

Nature conservation attributes of category III protected areas fall into two main types:

- Biodiversity that is uniquely related to the ecological conditions associated with the natural feature – such as the spray zones of a waterfall, the ecological conditions in caves or plant species confined to cliffs.
- Biodiversity that is surviving because the presence of cultural or spiritual values at the site have maintained a natural or semi-natural habitat in what is otherwise a modified ecosystem – such as some sacred natural sites or historical sites that have associated natural areas. In these cases the key criteria for inclusion as a protected area will be (i) value of the site as a contribution to broad-scale conservation and (ii) prioritization of biodiversity conservation within management plans.

Category III has been suggested as providing a natural management approach for many sacred natural sites, such as sacred groves. Although sacred natural sites are found in all categories and can benefit from a wide range of management approaches, they may be particularly suited to management as natural monuments.

Role in the landscape/seascape

Category III is really intended to protect the unusual rather than to provide logical components in a broad-scale approach to conservation, so that their role in landscape or ecoregional strategies may sometimes be opportunistic rather than planned. In other cases (e.g., cave systems) such sites may play a key ecological role identified within wider conservation plans:

- Important natural monuments can sometimes provide an *incentive* for protection and an *opportunity* for environmental/cultural education even in areas where other forms of protection are resisted due to population or development pressure, such as important sacred or cultural sites and in

these cases category III can preserve samples of natural habitat in otherwise cultural or fragmented landscapes.

What makes category III unique?

Because it is aimed at protecting a particular feature, category III is perhaps the most heavily influenced of all the categories by human perceptions of what is of value in a landscape or seascape rather than by any more quantitative assessments of value. This is less applicable in category III protected areas designated for geological features, where systematic identification is possible. Management is usually focused on protecting and maintaining particular natural features.

The fact that an area contains an important natural monument does not mean that it will inevitably be managed as a category III; for instance the Grand Canyon in Arizona is managed as category II, despite being one of the most famous natural monuments in the world, because it is also a large and diverse area with associated recreation activities making it better suited to a category II model. Category III is most suitable where the protection of the feature is the sole or dominant objective.

Category III differs from the other categories in the following ways:	
Category Ia	Category III is not confined to natural and pristine landscapes but could be established in areas that are otherwise cultural or fragmented landscapes. Visitation and recreation is often encouraged and research and monitoring limited to the understanding and maintenance of a particular natural feature.
Category Ib	
Category II	The emphasis of category III management is not on protection of the whole ecosystem, but of particular natural features; otherwise category III is similar to category II and managed in much the same way but at a rather smaller scale in both size and complexity of management.
Category IV	The emphasis of category III management is not on protection of the key species or habitats, but of particular natural features.
Category V	Category III is not confined to cultural landscapes and management practices will probably focus more on stricter protection of the particular feature than in the case of category V.
Category VI	Category III is not aimed at sustainable resource use.

Issues for consideration

- It will sometimes be difficult to ascertain the conservation attributes of category III sites, particularly in cases where there may be pressure to accept sites within a protected area system to help protect cultural or spiritual values.

- Not all natural monuments are permanent – while some sacred trees have survived for a thousand years or more they will eventually die – indeed many trees are considered to be sacred in part because they are already very old. It is not clear what happens to a category III protected area if its key natural monument dies or degrades.
- It is sometimes difficult to draw the boundaries between a natural monument and cultural site, particularly where archaeological remains are included within category III.
- Some apparent "monuments" may require protection of a larger ecosystem to survive – for example a waterfall may require protection of a whole watershed to maintain its flow.

Category IV: Habitat/species management area

> **Category IV** protected areas aim to protect particular species or habitats and management reflects this priority. Many category IV protected areas will need regular, active interventions to address the requirements of particular species or to maintain habitats, but this is not a requirement of the category.

Before choosing a category, check first that the site meets the definition of a protected area (page 8).

Primary objective

- To maintain, conserve and restore species and habitats.[5]

Other objectives:

- To protect vegetation patterns or other biological features through traditional management approaches;
- To protect fragments of habitats as components of landscape or seascape-scale conservation strategies;
- To develop public education and appreciation of the species and/or habitats concerned;
- To provide a means by which the urban residents may obtain regular contact with nature.

Distinguishing features

Category IV protected areas usually help to protect, or restore: 1) flora species of international, national or local importance; 2) fauna species of international, national or local importance including resident or migratory fauna; and/or 3) habitats. The size of the area varies but can often be relatively small; this is however not a distinguishing feature. Management will differ depending on need. Protection may be sufficient to maintain particular habitats and/or species. However, as category IV protected areas often include *fragments* of an ecosystem, these areas may not be self-sustaining and will require regular and active management interventions to ensure the survival of

specific habitats and/or to meet the requirements of particular species. A number of approaches are suitable:

- ***Protection of particular species***: to protect particular target species, which will usually be under threat (e.g., one of the last remaining populations);
- ***Protection of habitats***: to maintain or restore habitats, which will often be fragments of ecosystems;
- ***Active management to maintain target species***: to maintain viable populations of particular species, which might include for example artificial habitat creation or maintenance (such as artificial reef creation), supplementary feeding or other active management systems;
- ***Active management of natural or semi-natural ecosystems***: to maintain natural or semi-natural habitats that are either too small or too profoundly altered to be self-sustaining, e.g., if natural herbivores are absent they may need to be replaced by livestock or manual cutting; or if hydrology has been altered this may necessitate artificial drainage or irrigation;
- **Active management of culturally-defined ecosystems**: to maintain cultural management systems where these have a unique associated biodiversity. Continual intervention is needed because the ecosystem has been created or at least substantially modified by management. The *primary aim* of management is maintenance of associated biodiversity.

Active management means that the overall functioning of the ecosystem is being modified by e.g., halting natural succession, providing supplementary food or artificially creating habitats: i.e., management will often include much more than just addressing threats, such as poaching or invasive species, as these activities take place in virtually all protected areas in any category and are therefore not diagnostic. Category IV protected areas will generally be publicly accessible.

Role in the landscape/seascape

Category IV protected areas frequently play a role in "plugging the gaps" in conservation strategies by protecting key species or habitats in ecosystems. They could, for instance, be used to:

- Protect critically endangered populations of species that need particular management interventions to ensure their continued survival;
- Protect rare or threatened habitats including fragments of habitats;
- Secure stepping-stones (places for migratory species to feed and rest) or breeding sites;
- Provide flexible management strategies and options in buffer zones around, or connectivity conservation corridors between, more strictly protected areas that are more acceptable to local communities and other stakeholders;

---

[5] This is a change from the 1994 guidelines, which defined Category IV as protected areas that need regular management interventions. The change has been made because this was the only category to be defined by the process of management rather than the final objective and because in doing so it meant that small reserves aimed to protect habitats or individual species tended to fall outside the categories system.

- Maintain species that have become dependent on cultural landscapes where their original habitats have disappeared or been altered.

What makes category IV unique?

Category IV provides a management approach used in areas that have already undergone substantial modification, necessitating protection of remaining fragments, with or without intervention.

Category IV differs from the other categories in the following ways:	
Category Ia	Category IV protected areas are not strictly protected from human use; scientific research may take place but generally as a secondary objective.
Category Ib	Category IV protected areas can not be described as "wilderness", as defined by IUCN. Many will be subject to management intervention that is inimical to the concept of category Ib wilderness areas; those that remain un-managed are likely to be too small to fulfil the aims of category Ib.
Category II	Category IV protected areas aim their conservation at particular species or habitats and may in consequence have to pay less attention to other elements of the ecosystem in consequence, whereas category II protected areas aim to conserve fully functional ecosystems. Categories II and IV may in some circumstances closely resemble each other and the distinction is partly a matter of objective – i.e., whether the aim is to protect to the extent possible the entire ecosystem (category II) or is focused to protect a few key species or habitats (category IV).
Category III	The objective of category IV areas is of a more biological nature whereas category III is site-specific and more morphologically or culturally oriented.
Category V	Category IV protected areas aim to protect identified target species and habitats whereas category V aims to protect overall landscapes/seascapes with value for nature conservation. Category V protected areas will generally possess socio-cultural characteristics that may be absent in IV. Where category IV areas may use traditional management approaches this will explicitly be to maintain associated species as part of a management plan and not more broadly as part of a management approach that includes a wide range of for-profit activities.
Category VI	Management interventions in category IV protected areas are primarily aimed at maintaining species or habitats while in category VI protected areas they are aimed at linking nature conservation with the sustainable use of resources. As with category V, category VI protected areas are generally larger than category IV.

Issues for consideration

- Many category IV protected areas exist in crowded landscapes and seascapes, where human pressure is comparatively greater, both in terms of potential illegal use and visitor pressure.
- The category IV protected areas that rely on regular management intervention need appropriate resources from the management authority and can be relatively expensive to maintain unless management is undertaken voluntarily by local communities or other actors.
- Because they usually protect part of an ecosystem, successful long-term management of category IV protected areas necessitates careful monitoring and an even greater-than-usual emphasis on overall ecosystem approaches and compatible management in other parts of the landscape or seascape.

Category V: Protected landscape/seascape

A protected area where the interaction of people and nature over time has produced an area of distinct character with significant ecological, biological, cultural and scenic value: and where safeguarding the integrity of this interaction is vital to protecting and sustaining the area and its associated nature conservation and other values.

Before choosing a category, check first that the site meets the definition of a protected area (page 8).

Primary objective

- To protect and sustain important landscapes/seascapes and the associated nature conservation and other values created by interactions with humans through traditional management practices.

Other objectives

- To maintain a balanced interaction of nature and culture through the protection of landscape and/or seascape and associated traditional management approaches, societies, cultures and spiritual values;
- To contribute to broad-scale conservation by maintaining species associated with cultural landscapes and/or by providing conservation opportunities in heavily used landscapes;
- To provide opportunities for enjoyment, well-being and socio-economic activity through recreation and tourism;
- To provide natural products and environmental services;
- To provide a framework to underpin active involvement by the community in the management of valued landscapes or seascapes and the natural and cultural heritage that they contain;

- To encourage the conservation of agrobiodiversity[6] and aquatic biodiversity;
- To act as models of sustainability so that lessons can be learnt for wider application.

Distinguishing features

Category V protected areas result from biotic, abiotic and human interaction and should have the following *essential* characteristics:

- Landscape and/or coastal and island seascape of high and/or distinct scenic quality and with significant associated habitats, flora and fauna and associated cultural features;
- A balanced interaction between people and nature that has endured over time and still has integrity, or where there is reasonable hope of restoring that integrity;
- Unique or traditional land-use patterns, e.g., as evidenced in sustainable agricultural and forestry systems and human settlements that have evolved in balance with their landscape.

The following are *desirable* characteristics:

- Opportunities for recreation and tourism consistent with life style and economic activities;
- Unique or traditional social organizations, as evidenced in local customs, livelihoods and beliefs;
- Recognition by artists of all kinds and in cultural traditions (now and in the past);
- Potential for ecological and/or landscape restoration.

Role in the landscape/seascape

Generally, category V protected areas play an important role in conservation at the landscape/seascape scale, particularly as part of a mosaic of management patterns, protected area designations and other conservation mechanisms:

- Some category V protected areas act as a buffer around a core of one or more strictly protected areas to help to ensure that land and water-use activities do not threaten their integrity;
- Category V protected areas may also act as linking habitat between several other protected areas.

Category V offers unique contributions to conservation of biological diversity. In particular:

- Species or habitats that have evolved in association with cultural management systems and can only survive if those management systems are maintained;
- To provide a framework when conservation objectives need to be met over a large area (e.g., for top predators) in crowded landscapes with a range of ownership patterns, governance models and land use;
- In addition, traditional systems of management are often associated with important components of agrobiodiversity or aquatic biodiversity, which can be conserved only by maintaining those systems.

What makes category V unique?

Category V differs from the other categories in the following ways:	
Category Ia	Human intervention is expected. Category V does not prioritize research, though it can offer opportunities to study interactions between people and nature.
Category Ib	Category V protected areas are not "wilderness" as defined by IUCN. Many will be subject to management intervention inimical to the concept of category Ib.
Category II	Category II seeks to minimize human activity in order to allow for "as natural a state as possible". Category V includes an option of continuous human interaction.
Category III	Category III focuses on specific features and single values and emphasises the monumentality, uniqueness and/or rarity of individual features, whereas these are not required for category V protected areas, which encompasses broader landscapes and multiple values.
Category IV	Category V aims to protect overall landscapes and seascapes that have value to biodiversity, whereas category IV aims often quite specifically to protect identified target species and habitats. Category V protected areas will often be larger than category IV.
Category VI	Category VI emphasises the need to link nature conservation in natural areas whilst supporting sustainable livelihoods: conversely category V emphasises values from long-term interactions of people and nature in modified conditions. In category VI the emphasis is on sustainable use of environmental products and services (typically hunting, grazing, management of natural resources), whereas in category V the emphasis is on more intensive uses (typically agriculture, forestry, tourism). Category VI will usually be more "natural" than category V.

Issues for consideration

- Being a relatively flexible model, category V may sometimes offer conservation options where more strictly protected areas are not feasible.
- Category V protected areas can seek to maintain current practices, restore historical management systems or, perhaps most commonly, maintain key landscape values whilst accommodating contemporary development and change: decisions about this need to be made in management plans.
- The emphasis on interactions of people and nature over time raises the conceptual question for any individual category V protected area: at what point on the temporal continuum

---

[6] See definition in the Appendix.

should management focus? And, in an area established to protect values based on traditional management systems, what happens when traditions change or are lost?

- Since social, economic and conservation considerations are all integral to the category V concept, defining measures of performance for all of these values is important in measuring success.

- As people are the stewards of the landscape or seascape in category V protected areas, clear guidelines are needed about the extent to which decision making can be left to local inhabitants and how far a wider public interest should prevail when there is conflict between local and national needs.

- How is category V distinguished from sustainable management in the wider landscape? As an area with exceptional values? As an example of best practice in management? Category V is perhaps the most quickly developing of any protected area management approaches.

- There are still only a few examples of the application of category V in coastal and marine settings where a "protected seascape" approach could be the most appropriate management option and more examples are needed (see e.g., Holdaway undated).

Category VI: Protected area with sustainable use of natural resources

> **Category VI** protected areas conserve ecosystems and habitats, together with associated cultural values and traditional natural resource management systems. They are generally large, with most of the area in a natural condition, where a proportion is under sustainable natural resource management and where low-level non-industrial use of natural resources compatible with nature conservation is seen as one of the main aims of the area.

Before choosing a category, check first that the site meets the definition of a protected area (page 8).

Primary objective

- To protect natural ecosystems and use natural resources sustainably, when conservation and sustainable use can be mutually beneficial.

Other objectives

- To promote sustainable use of natural resources, considering ecological, economic and social dimensions;
- To promote social and economic benefits to local communities where relevant;
- To facilitate inter-generational security for local communities' livelihoods – therefore ensuring that such livelihoods are sustainable;

- To integrate other cultural approaches, belief systems and world-views within a range of social and economic approaches to nature conservation;
- To contribute to developing and/or maintaining a more balanced relationship between humans and the rest of nature;
- To contribute to sustainable development at national, regional and local level (in the last case mainly to local communities and/or indigenous peoples depending on the protected natural resources);
- To facilitate scientific research and environmental monitoring, mainly related to the conservation and sustainable use of natural resources;
- To collaborate in the delivery of benefits to people, mostly local communities, living in or near to the designated protected area;
- To facilitate recreation and appropriate small-scale tourism.

Distinguishing features

- Category VI protected areas, uniquely amongst the IUCN categories system, have the sustainable use of natural resources as a *means* to achieve nature conservation, together and in synergy with other actions more common to the other categories, such as protection.
- Category VI protected areas aim to conserve ecosystems and habitats, together with associated cultural values and natural resource management systems. Therefore, this category of protected areas tends to be relatively large (although this is not obligatory).
- The category is not designed to accommodate large-scale industrial harvest.
- In general, IUCN recommends that a proportion of the area is retained in a natural condition,[7] which in some cases might imply its definition as a no-take management zone. Some countries have set this as two-thirds; IUCN recommends that decisions need to be made at a national level and sometimes even at the level of individual protected areas.

Role in the landscape/seascape

- Category VI protected areas are particularly adapted to the application of landscape approaches.
- This is an appropriate category for large natural areas, such as tropical forests, deserts and other arid lands, complex wetland systems, coastal and high seas, boreal forests etc. – not only by establishing large protected areas, but also by linking with groups of protected areas, corridors or ecological networks.
- Category VI protected areas may also be particularly appropriate to the conservation of natural ecosystems when there are few or no areas without use or occupation and where those uses and occupations are mostly traditional and low-impact practices, which have not substantially affected the natural state of the ecosystem.

---

[7] Note that this does not necessarily preclude low-level activity, such as collection of non-timber forest products.

What makes category VI unique?

Allocation of category VI depends on long-term management objectives and also on local specific characteristics. The following table outlines some of the main reasons why category VI may be chosen in specific situations *vis-à-vis* other categories.

Category VI differs from the other categories in the following ways:	
Category Ia	Category VI protected areas do conserve biodiversity, particularly at ecosystem and landscape scale, but the aim would not be to protect them strictly from human interference. Although scientific research may be important, it would be considered a priority only when applied to sustainable uses of natural resources, either in order to improve them, or to understand how to minimize the risks to ecological sustainability.
Category Ib	Category VI protected areas in certain cases could be considered close to "wilderness", however they explicitly promote sustainable use, unlike the situation in category Ib wilderness areas where such use will be minimal and incidental to conservation aims. They also contribute to the maintenance of environmental services, but not only by exclusive nature conservation, as the sustainable use of natural resources can also contribute to the protection of ecosystems, large habitats, and ecological processes.
Category II	Category VI protected areas aim to conserve ecosystems, as complete and functional as possible, and their species and genetic diversity and associated environmental services, but differ from category II in the role they play in the promotion of sustainable use of natural resources. Tourism can be developed in category VI protected areas, but only as a very secondary activity or when they are part of the local communities' socio-economic strategies (e.g., in relation to ecotourism development).
Category III	Category VI protected areas might include the protection of specific natural or cultural features, including species and genetic diversity, among their objectives, whenever the sustainable use of natural resources is also part of the objectives, but they are more oriented to the protection of ecosystems, ecological processes, and maintenance of environmental services through nature protection and promotion of management approaches that lead to the sustainable use of natural resources.
Category IV	Category VI protected areas are more oriented to the protection of ecosystems, ecological processes, and maintenance of environmental services through nature protection and promotion of the sustainable use of natural resources. While category IV protected areas tend to prioritize active management, category VI promotes the sustainable use of natural resources.
Category V	Category V applies to areas where landscapes have been transformed as a result of long-term interactions with humans; category VI areas remain as predominantly natural ecosystems. The emphasis in category VI is therefore more on the protection of natural ecosystems and ecological processes, through nature protection and promotion of the sustainable use of natural resources.

Issues for consideration

- Protection of natural ecosystems and promotion of sustainable use must be integrated and mutually beneficial; category VI can potentially demonstrate best management practices that can be more widely used.
- New skills and tools need to be developed by management authorities to address the new challenges that emerge from planning, monitoring and managing sustainable use areas.
- There is also need for development of appropriate forms of governance suitable for category VI protected areas and the multiple stakeholders that are often involved. Landscape-scale conservation inevitably includes a diverse stakeholder group, demanding careful institutional arrangements and approaches to innovative governance.

Relationship between the categories

- The categories do not imply a simple hierarchy in terms of quality, importance or naturalness

- Nor are the categories necessarily equal in each situation, but rather should be chosen in order to maximize opportunities for conservation and also to address threats to conservation

The categories do not imply a simple hierarchy, either in terms of quality and importance or in other ways – for example the degree of intervention or naturalness. But nor are all categories equal in the sense that they will all be equally useful in any situation. One of the associated principles to the protected area definition states: "*All categories make a contribution to conservation but objectives should be chosen with respect to the particular situation; not all categories are equally useful in every situation*".

This implies that a well-balanced protected area system should consider using all the categories, although it may not be the case that all of the options are necessary or practical in every region or country. In the large majority of situations, at least a proportion of protected areas should be in the more strictly protected categories i.e., I–IV. Choice of categories is often a complex challenge and should be guided by the needs and urgency of biodiversity conservation, the opportunities for delivery of ecosystems services, the needs, wants and beliefs of human communities, land ownership patterns, strength of governance and population levels. Decisions relating to protected areas will usually be subject to a certain amount of trade-offs as a result of competing land uses and of consultative processes. It is important that conservation objectives are given adequate attention and weight in relevant decision-making processes.

Management approaches and categories are not necessarily fixed forever and can and do change if conditions change or if one approach is perceived to be failing; however changing the category of a protected area should be subject to procedures that are at least as rigorous as those involved in the establishment of the protected area and its category in the first place.

Many people assume that the categories imply a gradation in naturalness in order from I to VI but the reality is more complicated as shown in Figure 1 below, which attempts to compare average naturalness of all the categories.

Figure 1.   Naturalness and IUCN protected area categories

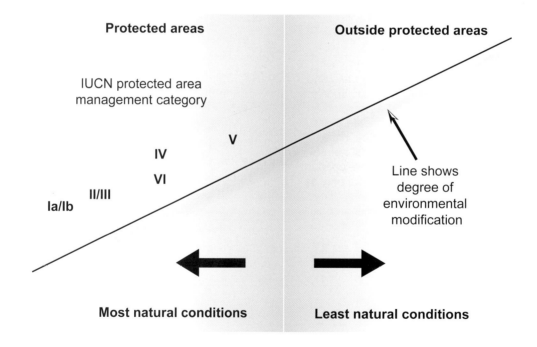

3. Governance

Categories are independent of who owns, controls, or has responsibility for management. However, governance is also very important. IUCN has identified diverse governance types in order to help in understanding, planning for and recording protected areas. This section outlines the IUCN governance types, explains how they link to the categories and looks at how governance by indigenous peoples, communities and private bodies can contribute to protected area systems.

Governance of protected areas

> IUCN recognises four broad types of governance of protected areas, any of which can be associated with any management objective:
>
> A. Governance by government
> B. Shared governance
> C. Private governance
> D. Governance by indigenous peoples and local communities

The IUCN protected area definition and management categories are "neutral" about types of ownership or management authority. In other words, the land, water and natural resources in any management category can be owned and/or directly managed by governmental agencies, NGOs, communities, indigenous peoples and private parties – alone or in combination. Both IUCN and the CBD recognise the legitimacy of a range of governance types. With respect to who holds decision-making and management authority and responsibility about protected areas, IUCN distinguishes four broad protected area governance types:

Type A: Governance by government (at federal/state/sub-national or municipal level). A government body (such as a Ministry or Park Agency reporting directly to the government) holds the authority, responsibility and accountability for managing the protected area, determines its conservation objectives (such as the ones that distinguish the IUCN categories), develops and enforces its management plan and often also owns the protected area's land, water and related resources. Sub-national and municipal government bodies can also be in charge of the above and/or own land and resources in protected areas. In some cases, the government retains the control of a protected area – in other words decides the *objectives* of managing the area – but delegates the planning and/or daily management tasks to a para-statal organization, NGO, private operator or community. Under a state's legal framework and governance there may or may not be a legal obligation to inform or consult stakeholders prior to setting up protected areas and making or enforcing management decisions. Participatory approaches are however increasingly common and generally desirable. Accountability measures also vary according to the country.

Type B: Shared governance. Complex institutional mechanisms and processes are employed to share management authority and responsibility among a plurality of (formally and informally) entitled governmental and non-governmental actors. Shared governance, sometimes also referred to as co-management, comes in many forms. In "collaborative" management, decision-making authority and responsibility rest with one agency but the agency is required – by law or policy – to inform or consult other stakeholders. Participation in collaborative management can be strengthened by assigning to multi-stakeholder bodies the responsibility of developing technical proposals for protected area regulation and management, to be submitted ultimately to a decision-making authority for approval. In "joint" management, various actors

sit on a management body with decision-making authority and responsibility. Decisions may or may not require consensus. In any of these cases, once decisions about management are taken, their implementation needs to be delegated to agreed bodies or individuals. One particular form of shared governance relates to transboundary protected areas, which involve at least two or more governments and possibly other local actors.

Type C: Private governance. Private governance comprises protected areas under individual, cooperative, NGO or corporate control and/or ownership, and managed under not-for-profit or for-profit schemes. Typical examples are areas acquired by NGOs explicitly for conservation. Many individual landowners also pursue conservation out of respect for the land and a desire to maintain its aesthetic and ecological values. Incentive schemes, such as revenues from ecotourism and hunting or the reduction of levies and taxes, often support this governance type. In all these cases, the authority for managing the protected land and resources rests with the landowners, who determine the conservation objective, develop and enforce management plans and remain in charge of decisions, subject to applicable legislation. In cases where there is no official recognition by the government, the accountability of private protected areas to society may be limited. Some accountability, for example in terms of long-term security, can be negotiated with the government in exchange for specific incentives (as in the case of Easements or Land Trusts).

Type D: Governance by indigenous peoples and local communities. This type includes two main subsets: (1) indigenous peoples' areas and territories established and run by indigenous peoples and (2) community conserved areas established and run by local communities. The subsets, which may not be neatly separated, apply to both sedentary and mobile peoples and communities. IUCN defines this governance type as: *protected areas where the management authority and responsibility rest with indigenous peoples and/or local communities through various forms of customary or legal, formal or informal, institutions and rules.* These can be relatively complex. For instance, land and/or sea resources may be collectively owned and managed while other resources may be managed individually or on a clan basis. Different indigenous peoples or communities may be in charge of the same area at different times, or of different resources within the same area. Rules generally intertwine with cultural and spiritual values. The customary rules and organizations managing natural resources often possess no statutory legal recognition or sanctioning power. In other cases, however, indigenous peoples and/or local communities are fully recognised as the legitimate authority in charge of state-listed protected areas or have legal title to the land, water or resources. Whatever the structure, the governance arrangements require that the area under the control of indigenous peoples and/or local communities has identifiable institutions and regulations that are responsible for achieving the protected area objectives.

The four governance types outlined above are taken into consideration together with the management categories in the following matrix (adapted from Borrini-Feyerabend *et al.* 2004).

Table 3. **"The IUCN protected area matrix": a classification system for protected areas comprising both management category and governance type**

Governance types / Protected area categories	A. Governance by government			B. Shared governance			C. Private governance			D. Governance by indigenous peoples and local communities	
	Federal or national ministry or agency in charge	Sub-national ministry or agency in charge	Government-delegated management (e.g., to an NGO)	Transboundary management	Collaborative management (various forms of pluralist influence)	Joint management (pluralist management board)	Declared and run by individual land-owners	... by non-profit organizations (e.g., NGOs, universities)	... by for-profit organizations (e.g., corporate owners, cooperatives)	Indigenous peoples' protected areas and territories – established and run by indigenous peoples	Community conserved areas – declared and run by local communities
Ia. Strict Nature Reserve											
Ib. Wilderness Area											
II. National Park											
III. Natural Monument											
IV. Habitat/ Species Management											
V. Protected Landscape/ Seascape											
VI. Protected Area with Sustainable Use of Natural Resources											

Note that governance types describe the different types of management authority and responsibility that can exist for protected areas but do not necessarily relate to ownership. In some of the governance types – e.g., state and private protected areas – governance and ownership will often be the same. However in other cases this will depend on individual country legislation: for example many indigenous peoples' protected areas and community conserved areas are found on state-owned land. In large and complex protected areas, particularly in categories V and VI, there may be multiple governance types within the boundaries of one protected area, possibly under the umbrella of an overview authority. In the case of most marine protected areas the ownership can be with the state, which will either manage directly or delegate management to communities, NGOs or others. There are, however, many marine areas where the customary laws of indigenous peoples are recognised and respected by the broader society. In international waters and the Antarctic, where there is no single state authority, protected areas will inevitably need to be under a shared governance type.

Recording governance types

IUCN suggests that the governance type of a protected area be identified and recorded at the same time as its management objective (category) in national environmental statistics and accounting systems and in protected area databases. In some cases deciding on the governance type may be as or more delicate and complex than identifying the category and one may inform and influence the other; also, many protected areas are likely to change their governance types over time. As mentioned, in the case of large protected areas, several governance types may exist within the boundary of a single area.

In considering governance for the purpose of reporting to the World Database on Protected Areas, IUCN WCPA proposes adopting a two-dimensional structure. Though management objectives for the categories can be developed and assigned without regard for governance, comparisons of protected areas and their effectiveness will be greatly enhanced by listing governance type as well as management category in future databases. The protected area categories are not taxonomic, unlike the governance types; however, a two-dimensional classification can easily sort for both management objectives (i.e., category I–VI) and governance type (i.e., A–D, as described above). Using the letter designations used above, for example, Yellowstone National Park (USA) might be described as category II-A; Mornington Wildlife Sanctuary (Australia) might be II-C; Snowdonia National Park (UK) V-B; and Coron Island (The Philippines) as a combination of I-D and V-D.

Governance quality

For protected areas in all management categories, management effectiveness provides a measure of the actual achievement of the conservation goals. Management effectiveness is also influenced by governance quality, that is, "how well" a governance regime is functioning. In other words, the concept of governance quality applied to any specific situation attempts to provide answers to questions such as "Is this 'good' governance? and "Can this governance setting be 'improved' to achieve both conservation and livelihood benefits?"

"Good governance of a protected area" can be understood as a governance system that responds to the principles and values freely chosen by the concerned people and country and enshrined in their constitution, natural resource law, protected area legislation and policies and or cultural practices and customary laws. These should reflect internationally agreed principles for good governance (e.g., Graham *et al.* 2003). International agreements and instruments have set governance principles and values, such as the CBD, the Aarhus Convention, the UN Convention to Combat Desertification, the Universal Declaration of Human Rights and the UN Declaration on the Rights of Indigenous Peoples. A number of international and regional processes have also been critical in setting this agenda, including the 2003 World Parks Congress in South Africa, the 2005 First Congress of Marine Protected Areas in Australia and the 2007 Second Latin American Protected Areas Congress in Argentina. Drawing from these and field experience IUCN has explored a set of broad principles for good governance of protected areas, including:

- **Legitimacy and voice** – social dialogue and collective agreements on protected area management

objectives and strategies on the basis of freedom of association and speech with no discrimination related to gender, ethnicity, lifestyles, cultural values or other characteristics;
- **Subsidiarity** – attributing management authority and responsibility to the institutions closest to the resources at stake;
- **Fairness** – sharing equitably the costs and benefits of establishing and managing protected areas and providing a recourse to impartial judgement in case of related conflict;
- **Do no harm** – making sure that the costs of establishing and managing protected areas do not create or aggravate poverty and vulnerability;
- **Direction** – fostering and maintaining an inspiring and consistent long-term vision for the protected area and its conservation objectives;
- **Performance** – effectively conserving biodiversity whilst responding to the concerns of stakeholders and making a wise use of resources;
- **Accountability** – having clearly demarcated lines of responsibility and ensuring adequate reporting and answerability from all stakeholders about the fulfilment of their responsibilities;
- **Transparency** – ensuring that all relevant information is available to all stakeholders;
- **Human rights** – respecting human rights in the context of protected area governance, including the rights of future generations.

Governance by indigenous peoples and local communities, and private governance are discussed in greater detail below.

Governance by indigenous peoples and local communities

A note on terminology: concepts of governance by indigenous peoples and local communities are still evolving and differ around the world. Some indigenous peoples wish to see their territories clearly distinguished from those of local communities. In other cases, indigenous peoples and local communities are co-inhabiting and co-managing areas, and in yet further cases indigenous peoples use the term "community conserved areas" for practical reasons, for example when the term "indigenous" is not recognised. Similar regional differences exist regarding the term "territory". Amongst both indigenous peoples and local communities there are cases where the term "conserved area" is used and others where "protected area" is preferred: we use a range of terms herein. Below we summarise the concepts and include a description of indigenous peoples' territories and protected areas.

Although some of the protected areas governed by indigenous peoples and local communities have been in existence for hundreds or even thousands of years, their recognition by national governments and their inclusion within national protected area systems is a much more recent phenomenon, which deserves particular attention here. Indigenous peoples' protected areas, indigenous peoples' conserved territories and community conserved areas (which we summarise as Indigenous and community conserved areas or ICCAs) have three essential characteristics:

- The relevant indigenous peoples and/or local communities are closely concerned about the relevant ecosystems – usually being related to them culturally (e.g., because of their value as sacred areas) and/or because they support their livelihoods, and/or because they are their traditional territories under customary law.
- Such indigenous peoples and/or local communities are the major players ("hold power") in decision making and implementation of decisions on the management of the ecosystems at stake, implying that they possess an institution exercising authority and responsibility and capable of enforcing regulations.
- The management decisions and efforts of indigenous peoples and/or local communities lead and contribute towards the conservation of habitats, species, ecological functions and associated cultural values, although the original intention might have been related to a variety of objectives, not necessarily directly related to the protection of biodiversity.

There is mounting evidence that ICCAs that meet the protected area definition and standards can provide effective biodiversity conservation responding to any of the management objectives of the IUCN categories, and particularly so in places where protected areas governed by government are politically or socially impossible to implement or likely to be poorly managed. ICCAs are starting to be recognised as part of conservation planning strategies, complementing government-managed protected areas, private protected areas and various forms of shared governance (see http://www.iccaforum.org/). But this is still more the exception than the rule.

Most ICCAs are at present not formally recognised, protected or even valued as part of national protected area systems. In some cases, there may be good reasons for this – including reluctance of the relevant indigenous peoples and/or local communities to becoming better known or disturbed, for instance when the site has sacred values that require privacy or when the relevant indigenous peoples choose to manage their land in accordance with customary laws only. As countries move towards greater recognition of ICCAs, these sensitivities need to be kept in mind. Depending on the specific situation and the main concerns of the relevant indigenous peoples or local communities, appropriate government responses may vary from incorporation of the ICCA into the national protected area system, to recognition "outside of the system", to no formal recognition whatsoever. This last option, of course, should be selected when formal recognition may undermine or disturb the relevant ICCAs.

Most ICCAs face formidable forces of change, which they might be better able to withstand with the help of an official recognition and appreciation, especially when the most likely alternative may be exploitation, e.g., for timber or tourism. In these cases recognition within national protected area systems, if ICCAs meet the protected area definition and standards or other types of formal recognition, can provide indigenous peoples and local communities with additional safeguards over their land. This should be coupled, however, with the acceptance by the state that ICCAs are inherently different from state-governed protected areas – in particular regarding their governing institutions. It should be noted however that formal recognition of ICCAs can bring new dangers, such as increased visitation and commercial attention to the site, or greater governmental interference. Indigenous peoples and local communities also worry that official recognition of ICCAs may get them co-opted into larger systems over which they have, basically, no control.

Although there is growing recognition of the positive role that ICCAs can play in maintaining biodiversity, there is also concern in the conservation community that "weak" ICCAs could be added to national protected area systems as a cheaper and more politically-expedient alternative to other conservation options. There are also worries that, as societies change, community approaches to management may also change and some of the traditional values and attitudes that helped in conserving biodiversity might be lost in the process. Formal ICCAs that are unable to maintain their traditional conservation practices are worse than informal, unrecognised ICCAs.

Ultimately, and bearing in mind all the cautionary issues mentioned above, the recognition of ICCAs that fully meet protected area definitions and standards in national and regional protected area strategies is one of the most important contemporary developments in conservation. Some initial thinking on the criteria for recognition has already been published (Borrini-Feyerabend *et al.* 2004) and further developments are expected as part of the IUCN/WCPA Best Practice Guidelines for Protected Areas series.

Indigenous peoples' territories and protected areas

Especially in regions such as Latin America, North America, Oceania, Africa, Asia and the Arctic, many formally designated protected areas are at the same time the ancestral lands and waters of indigenous peoples, cultures and communities. IUCN has long adopted and promoted protected area policies that respect the rights and interests of indigenous peoples, and has developed tools and approaches to facilitate their recognition and implementation.

Consistent with its policies, IUCN applies the following principles of good governance as they relate to protected areas overlapping with indigenous peoples' traditional lands, waters and resources:

- Protected areas established on indigenous lands, territories and resources should respect the rights of traditional owners, custodians, or users to such lands, territories and resources;

- Protected area management should also respect indigenous peoples' institutions and customary laws;

- Therefore protected areas should recognise indigenous owners or custodians as holders of the statutory powers in their areas, and therefore respect and strengthen indigenous peoples' exercising of authority and control of such areas.

In recent years there have been many important developments in relation to protected areas overlapping with indigenous peoples' lands, waters and resources. First, IUCN at its World Conservation Congresses has adopted specific policies on protected areas and indigenous peoples' rights. Secondly, at the national level many countries have adopted and applied new legal and policy frameworks relevant to indigenous peoples' rights, with important implications for protected areas. At the international level, several instruments such as the CBD *Programme of Work on Protected Areas*, as well as the UN Declaration on the Rights of Indigenous Peoples, have been adopted and have changed significantly the political landscape regarding indigenous peoples and protected areas.

Following such policy developments, important changes have also occurred on the ground. Many state-declared protected areas overlapping with indigenous peoples' lands, waters and resources have entered into shared governance arrangements and moved towards self-management by indigenous peoples. In countries like Canada, Australia, New Zealand, and several countries in Latin America, many new protected areas have been created at the request or initiative of indigenous owners, or through joint arrangements with governments. In such cases, indigenous land and resource rights, as well as indigenous government of the land, are key features.

Many indigenous peoples see protected areas as a very useful tool for them, since they can strengthen protection of their territories, lands and resources against external threats, offer new opportunities for sustainable use, strengthen culture-based protection of critical places, and consolidate indigenous institutions for land management. In such conditions, indigenous peoples' protected areas are a growing and important phenomenon, and one that is likely to increase around the world.

Not all indigenous lands, territories and resources fully comply with the protected area definition, but some certainly do and can be considered as "protected areas". Accordingly, indigenous peoples' protected areas can be defined as:

"clearly defined geographical spaces, within the lands and waters under traditional occupation and use by a given indigenous people, nation or community, that are voluntarily dedicated and managed, through legal or other effective means including their customary law and institutions, to achieve the long-term conservation of nature with associated ecosystem services, as well as the protection of the inhabiting communities and their culture, livelihoods and cultural creations".

The main distinguishing features of indigenous peoples' protected areas have to do with the socio-political arrangements that are established between indigenous peoples and national authorities for the government of lands and resources in indigenous peoples' lands. Basically such features are that:

1. They are based upon the collective rights of the respective indigenous people, nation or community to lands, territories and resources, under national contexts;

2. They are established as protected areas in application of the right of self-determination, exercised mainly through:

- Self-declaration of the protected area by the indigenous people or nation with collective territorial rights on the area;

- Free, prior and informed consent of the people, nation or community with territorial rights on the area, in cases where the designation proposal is originated in government agencies, conservation organizations or other actors.

3. They are based on ancestral or traditional occupation;

4. Occupation, use and management are connected to and dependent upon the broader socio-cultural and political structure of a people or nation, which includes their customary law and institutions;

5. They are self-governed by indigenous institutions within their territories and the protected areas contained therein, in application of arrangements established with system-level protected area authorities.

IUCN recognises that there should be specific guidance developed on the whole issue of indigenous peoples' territories and protected areas and hopes to be working with indigenous peoples' organizations around the world to make this a reality.

Possible steps to determine whether an indigenous peoples' territory or ICCA is a "protected area" and to recognise it in a national protected area system

- Determine whether the area and its current governance system fits within the protected area definition of IUCN.

- Determine whether the area also meets the criteria of a protected area under national legislation and policy.

- If so, determine whether it fits within the existing typology of protected area categories of the country concerned. Could the area qualify as a national park, sanctuary, game reserve, or other existing category? Importantly, would such a category allow for the community's own governance system to continue? Would it allow for management objectives that may be conceptually and/or practically different from conservation *per se*?

- When national legislation and policies are fully compatible with local practice, conservation agencies should grant, or formally recognise, that authority and decision-making powers for the establishment and management of the area should rest with the concerned indigenous peoples and/or local communities. Importantly, a fact which will directly enable them to enforce their decisions (as in the case in which an ordinance for the control of fishing may provide the needed legal backing to a community-declared marine sanctuary).

- When there is incompatibility between indigenous peoples or community governance of a valuable area and national protected area laws and regulations, legal and policy adjustments might be required to the current statutory provisions so that the relevant indigenous peoples and local communities can retain their governance systems. Often, what the indigenous peoples or local communities request is a guarantee of customary tenure, use and access rights sanctioned through a demarcation of territories and resources. For that to happen, however, it may be necessary that the institution governing the area be recognised as a legal body. As this can affect the ways indigenous peoples and local communities organize themselves and manage their areas and territories, it is important that they determine such matters.

- After incompatibilities are removed, the agency may embark on a process of negotiation, which may end in a contractual arrangement between concerned indigenous peoples and/or local communities and national or sub-national authorities. Such a contractual arrangement could, for instance, recognise the area and provide to it some form of legal protection or technical and financial support, including inclusion as an autonomous part of a national protected area system. In other cases, it may transform the area into a protected area under shared governance.

- Once agreement has been reached between the concerned indigenous peoples and/or local communities and national or sub-national authorities about recognising the area as a protected area, the relevant rules and regulations may need to be clarified and made public. This may involve the mere recording of existing customary rules, without interference from the state agencies, or the incorporation of new advice, methods and tools into these rules. The rules should specify what kind of land and resource zoning exist, what community and individual rights (including ownership) exist, what institutional structures manage the area, whether and how sustainable resource harvesting is allowed to take place (e.g., with limits on quantity, species and seasons) and what processes should be followed to de-recognise the area if its agreed conservation objectives are not being met. It may also be useful to clarify and record the subdivision of rights and responsibilities among the concerned indigenous peoples and local communities themselves and to specify provisions against the misuse of rights and power on the part of authorities at all levels.

- As part of the governance process, boundaries are to be effectively enforced and protected against external threats. What kind of customary and local surveillance and enforcement mechanisms are recognised by the state? For instance, can members of the concerned indigenous peoples and local communities apprehend violators? Is government help needed? Who judges in the event of controversies? Who is responsible for the information campaigns needed for the general public to respect ICCAs and indigenous protected areas? The answers to these questions are important for such areas to remain effective as protected areas through time.

Private governance

Private protected areas are a large and growing subset of the world's protected areas that have representatives in all the IUCN categories, but have until now been under-represented in the body of areas recognised by IUCN and reported in the WDPA.

Private protected areas are generally not under direct governmental authority. There are three entities in charge of private protected areas, each with particular management implications:

- Individual (the area is under control of a single person or family).
- NGO (the area is under control of a charitable not-for-profit organization operating to advance a specific

mission and usually controlled by an executive, a board and subscribing members). In rare cases this can include cooperatives (e.g., the Ahuenco Conservation Community in Chile).

- Corporate (the area is under the control of a private, for-profit company or group of people authorized to act as a single entity, usually controlled by an executive, an oversight board, and ultimately individual shareholders).

Each of these general sub-types (and myriad variations) has particular management implications. Indigenous peoples and local communities can also be formal owners and/or in control of land and resources they wish to conserve. Their case has just been discussed above.

Private protected areas in the categories

Private protected areas can and do fall into all the categories. Some people assume that they are better represented under categories IV–VI; but in fact many fit the management objectives of I–III, perhaps especially those owned/managed by NGOs. Although most marine waters are not privately owned, an increasing number of privately-owned islands are being protected, including their coastal and marine areas.

Most private protected areas are currently not recorded on the WDPA and are therefore largely unrecognised by the global community: they are also often effectively ignored by governments and not included within national or ecoregional planning. This may reflect a lack of governmental capacity to collect data on private protected areas, or private protected area managers/owners being reluctant to share information freely.

"Effective means"

In the majority of cases, the creation of a private protected area – and management of the same for conservation objectives – is a voluntary act on the part of the landowners. A growing recognition of the opportunities for achieving conservation objectives on private land – and especially the proliferation of mechanisms and incentives for doing so – has resulted in a dramatic increase in the number and extent of private protected areas. These mechanisms and incentives include:

- Systems of voluntary protected area designations, in which landowners agree to certain management objectives or restrictions in return for assistance or other incentives: the Private Natural Heritage Reserves of Brazil are an example.
- Voluntary surrender of legal rights to land use on private property, sometimes to realize advantages (for example in neighbouring land) conferred by the theoretical loss

in value, or to secure protection in perpetuity, or as compensation measures: mechanisms include conservation easements and related covenants and servitudes; and conservation management agreements.

- Charitable contributions, where NGOs raise funds privately or publicly for the purchase of land for protection, or receive gifts of land directly from willing donors: this includes large international NGOs such as The Nature Conservancy and Conservation International along with many national and local examples.
- Corporate set-aside, donations, or management of an area for conservation, stimulated by a desire for good public relations; as a concession or off-set for other activities; because it is stipulated in "green" certification; as an investment in the future; or due to personal interest of staff.
- Involuntary surrender of some management rights in response to legal restrictions.

The categories system holds the potential to assist governments in monitoring private conservation activities, through evaluating both the management objectives of private protected areas and their effectiveness. There are in addition local and national safeguards in place in some countries to ensure that private protected areas are managed according to designation, regulation or proclamation. The practical significance and implementation of these safeguards vary widely among countries. (There are also examples of self-regulation of private protected areas, such as the developing land trust accreditation programme in the United States). Application of the IUCN categories system set out in these guidelines could provide governments with a comparative basis for monitoring private protected areas within their national conservation strategies.

The IUCN definition of a protected area is clear that such areas should be managed for conservation in perpetuity and this is the main criterion that will distinguish whether a particular area of privately-owned land or water is or is not a protected area. A land owner who manages for conservation today but makes no provisions for whether or not the management will continue into the future is certainly contributing to conservation but not through a recognised protected area. Providing long-term security is one of the challenges facing private protected areas. Some national governments have addressed this through introducing legislation that makes declaration of a private protected area a legally-binding commitment over time although where this is not the case, other mechanisms may be necessary. These are still being developed and include various certification systems, institutionalized systems of declaration and peer pressure. Further work on steps needed to integrate private protected areas more fully into national and international protected area systems is urgently needed.

4. Applying the categories

This section describes the processes for applying categories, including: choosing and then agreeing the most suitable category for a given situation; assigning the category to meet national legal requirements and international standards and norms; and recording the protected area and category with the UNEP World Conservation Monitoring Centre. Questions about verifying categories and addressing disputes are also discussed.

Choosing the correct category

Once an area has been identified as a protected area according to the IUCN definition, the next stage in classification is to determine which category matches most closely the overall management objectives of the protected area.

As the categories system reflects management objectives, it follows that once a decision has been made about the management of a protected area the correct category should be obvious. This is sometimes how it happens. Unfortunately, in many other cases there is also plenty of room for confusion: perhaps because there are multiple objectives within a protected area (maybe in different parts of the area); or because protected area objectives are evolving and are often becoming more complex; or because there is still uncertainty about what particular approach works best. Agreeing objectives (perhaps reassessing the original objectives) and developing management plans are both closely linked to agreement of a category.

Many people have asked IUCN for a foolproof way of identifying a category but this is difficult. There are often several ways to approach management in the same protected area, which can therefore be categorized in different ways. What happens if most of a protected area is managed in one way but part of it in another? Is there a minimum size or maximum size for particular categories? Are international designations such as World Heritage or Ramsar associated with particular categories? How much human activity is "allowed" in protected areas in different categories? The following section attempts to answer these questions.

It should be remembered that many countries have legislation setting out clearly the criteria under which different types of protected areas are identified: these may or may not equate with the IUCN categories. In the latter case, countries that want to list their protected areas correctly on the WDPA need to work out the relationship between their own classification system and the IUCN categories – many have already done so. In other cases governments have taken the IUCN categories and further refined them for the specific conditions in the country. As long as the refining process does not undermine the basic principles of a protected area or of specific categories, IUCN encourages such a process. It follows that choice of category will vary with conditions and from one country to the next and can on occasion be a complicated process – as much art as science.

But before jumping into the technical details of the application of protected area categories it is also worth considering *why* categories are being chosen. Categorization can take place at three stages in the life of a protected area and although this should not influence the result, it may make important differences to the process. Categories can be selected:

- Before the protected area is established, when decisions about management objectives should be part of the planning process.
- After the protected area has been established, when management objectives have already been decided and choosing the appropriate category is mainly about finding the one that best fits the protected areas as a whole ; although looking carefully at the categories at this stage might also stimulate some changes in management objectives and activities.
- In an established protected area where there is already a category but either management is changing to address emerging conservation priorities and problems or there are doubts about whether the right category was chosen in the first place. However, changing a category in most countries is governed by the legal framework on protected areas and should follow an assessment at least as rigorous as the one applied in defining the existing category in the first place.

How does the management objective relate to the category?

- The category should be based on the primary management objective(s) of the protected area

- The primary management objective should apply to at least three-quarters of the protected area

THE CATEGORY SHOULD BE BASED AROUND THE PRIMARY MANAGEMENT OBJECTIVE(S): as listed for individual categories in Chapter 2. (It also needs to fit the definition of a protected area). This assumes that the agency responsible for the protected area is able to decide on the main aim of management. This is not necessarily an easy choice to make; on the other hand failure to do so suggests that management itself may be confused and likely to be ineffective. In principle a good assessment process to identify the right category should involve key stakeholders and other agencies dealing with the conservation and management of the protected area and should be based on best available natural and social science. Identifying a primary objective does not mean that other aims are not important: almost all protected areas have multiple values. In practice it is not always easy to make a judgement – the following examples look at some of the common questions that arise:

- *Ecosystem or habitat – category II or IV?* Category II protected areas are supposed to conserve whole ecosystems whereas category IV generally aims to conserve species or fragments of ecosystems. In fact, very few protected areas are large enough to protect entire ecosystems, with the associated migration routes, watershed functions etc. Distinguishing II and IV is therefore often a matter of degree: a category II protected area should aim to protect the majority of naturally-occurring ecosystem functions, while a category IV protected area is usually either a fragment of an ecosystem (e.g., a pool, fragment of coral reef or small area of bog) or an area that relies on regular management intervention to maintain an artificial ecosystem (e.g., a coppice woodland or regularly mown area of grassland). Category IV protected areas are generally smaller than category II although this is not diagnostic and large category IV protected areas exist.

- *Management intervention or cultural landscape – category IV or V?* A category IV protected area is managed primarily for its flora and fauna values, and interventions such as coppicing, vegetation clearance, prescribed burning etc. are undertaken mainly with this in mind: any profits or social benefits from such ventures are secondary. Management interventions in category V protected areas are conversely aimed at sustaining human livelihoods and are not just part of a biodiversity management strategy. A category V protected area therefore uses cultural management systems that also have a value for biodiversity, such as cork oak woodland that is managed primarily for cork but also has important wildlife values if integrated into a landscape approach to conservation. In most category V protected areas, a range of different management approaches are often combined.

- *Restoring a cultural landscape – category V or something else?* A cultural landscape would normally be category V. But if the aim of management is to restore a former cultural landscape into something much more natural, then the management objective and therefore in turn the category might fit better as something else, such as category Ib, or II or IV. For example, protecting relict woodland formerly used for sheep grazing with an aim to restoring it to something resembling the original forest ecosystem would not usually be classified as a category V protected area. Protecting a heavily exploited coral reef with the aim of restoring it back to a more pristine ecosystem would similarly not usually be classified as category V.

- *Natural monument or ecosystem – category III or II?* When is protection of a natural monument equivalent to protection of an ecosystem? In practice it is often a question of size and focus of management objectives. A protected area containing an important natural monument (normally category III), but nonetheless managed *primarily* for its ecosystem functions (normally category II) should be categorized as II rather than III – e.g., the Grand Canyon in Arizona is one of the largest natural monuments in the world but the national park is managed primarily for its ecosystem functions and is listed as II.

- *Sustainable use or incidental use by local communities – when to use category VI?* Many protected area categories permit limited human use; for example many wilderness areas (Ib) and protected ecosystems (II)

permit local people to carry out traditional small-scale livelihood activities that are in harmony with the nature in the protected area such as (depending on individual management agreements) reindeer herding, fishing, collection of non-timber forest products and limited subsistence hunting. But in these cases the objective is conservation of wilderness or ecosystems and human take-off should make a minimal impact on this. In category VI the objective of management is sustainable use in synergy with nature conservation and it is expected that the activities are managed in a way that does not produce a substantial impact on these ecosystems. The difference is partly a matter of degree.

- *Cultural landscape – what is not category V?* Few if any land areas have not been modified by human societies over hundreds or thousands (or tens of thousands) of years and many aquatic ecosystems have also been modified. It could be argued that every protected area in the world is a category V. But whilst recognising the role of human communities, IUCN distinguishes areas that have predominantly natural species and ecosystems (not usually category V) from those where the level of modification is more intense, such as areas with long-term settled farming or management processes that make major changes to ecology and species diversity (usually category V).

THE PRIMARY OBJECTIVE SHOULD APPLY TO AT LEAST THREE-QUARTERS OF THE PROTECTED AREA – THE 75 PERCENT RULE: many protected areas may have specific zones within them where other uses are permitted: e.g.:

- Tourist lodges and camps in category II national parks – as is the case with many African savannah protected areas;
- Villages remaining within otherwise strictly protected areas – e.g., a village remains within Cat Tien National Park in Viet Nam;
- Small strictly protected core areas in what is otherwise a cultural landscape managed as category V – e.g., woodlands owned by the National Trust in the Brecon Beacons National Park, Wales, UK;
- Areas where fishing is permitted within what is otherwise a strictly protected marine or freshwater protected area – e.g., in Kosi Bay Nature Reserve in KwaZulu Natal, South Africa.

IUCN recognises this and recommends that up to 25 percent of land or water within a protected area can be managed for other purposes so long as these are compatible with the primary objective of the protected area. In some cases, the 25 percent may be movable: for example Bwindi Impenetrable Forest National Park in Uganda permits local communities to collect medicinal plants and other non-timber forest products in specially designated zones that are moved occasionally to ensure that the species are not over-collected.

How is the category affected by size of protected area?

> • There are no hard and fast rules but some categories tend to be relatively larger or smaller

Overall scale often depends on other factors, such as the amount of land or water available, population density etc.

In terms of *relative scale* some categories are more likely to be either large or small, because of their particular management objectives, but there could be exceptions for virtually every category. To aid selection, Table 4 below suggests relative scale for the categories and explains why, but also gives some exceptions to show that size alone should not be a determining factor.

Table 4.   How size of protected area relates to the category

Cat.	Relative size	Explanation	Exceptions
Ia	Often small	Strictly protected, no-go areas are always difficult to agree except in sparsely inhabited areas: therefore although large Ia areas exist (e.g., in Australia) they are probably the exception.	Large areas in places with low human population density and little interest in tourism.
Ib	Usually large	Part of the rationale of wilderness areas is that they provide enough space to experience solitude and large-scale natural ecosystem.	Relatively small areas set up as wilderness in the hope that they can be expanded in the future.
II	Usually large	Conservation of ecosystem processes suggests that the area needs to be large enough to contain all or most such processes.	Small islands may effectively be ecosystems and thus functionally category II.
III	Usually small	Larger sites containing natural monuments would generally also protect other values (e.g., ecosystems and/or wilderness values).	
IV	Often small	If the site is set up to protect only individual species or habitats this suggests that it is relatively small.	Larger areas set aside as nature reserves but needing regular management to keep functioning might best be IV.
V	Usually large	The mosaic of different approaches adding up to conservation gains in landscape approaches suggests a larger area.	Some mini-reserves for crop wild relatives or land races may need cultural management.
VI	Usually large	The extensive nature of management suggests that it will usually be a large area.	Some marine category VI protected areas are small.

Can a protected area contain more than one category?

> • Distinct protected areas nested within larger protected areas can have their own category
>
> • Different zones in larger protected areas can have their own category, if the zones are described and fixed in law
>
> • Different protected areas making up a transboundary protected area may have different categories

This is one of the most vexed questions relating to the categories. The answer is that it depends; on ownership, governance and to some extent on the wishes of the protected area authority or authorities.

There are three situations where single or contiguous protected areas may be assigned different categories:

Nested areas with multiple objectives: protected areas of different categories are sometimes "nested" within another – i.e., a large protected area can contain several smaller protected areas inside. The most common model would be a large, less strictly protected area containing smaller, more strictly protected areas inside. For example, many category V areas contain within them category I and IV areas – possibly under completely different management authorities or governance approaches. The Vercors Regional Nature Park in France (category V) contains the Hauts Plateaux du Vercors within it (category IV). This is entirely consistent with the application of the categories system. When reporting "nested" protected areas it is important to avoid double counting and to ensure that databases do not overstate the amount of land or sea that has been designated. For example, in the UK, the national parks (category V), which cover about 9 percent of the land area of England and Wales, include a number of national nature reserves (category IV), covering about 0.7 percent of the area of the national parks.

Different zones within protected areas: zoning is usually a management tool within a single protected area and would not generally be identified by a separate category, but there are exceptions. In some protected areas, parts of a single management unit are classified *by law* as having different management objectives and being separate protected areas: in effect, these "parts" are individual protected areas that together make up a larger unit, although they are all under a single management authority. In the case of Australia, for example, zoning is used both as a management tool and as a tool for protected area designation and is enshrined in regulation. Thus the Great Barrier Reef Marine Park in Australia has been assigned category VI in its entirety, but has also been officially assigned other categories that relate to regulated management zones within the park. Separating zones into different categories is something that would usually only be attempted for large protected areas and is at the discretion of the government concerned, given the conditions described above.

IUCN recommends that multiple categories can be reported within a single large protected area when certain conditions are met. These conditions reflect the permanence and objectives of the zoning system. Two alternative scenarios are:

- **"Hard" zone:** zones can be assigned to an IUCN category when they: (a) are clearly mapped; (b) are recognised by legal or other effective means; and (c) have distinct and unambiguous management aims that can be assigned to a particular protected area category (the 75 percent rule is not relevant);
- **"Soft" zone:** zones are not assigned to an IUCN category when they: (a) are subject to regular review, such as through a management planning process; (b) are not recognised by legal or other effective means; and (c) do not correspond to a particular protected area category (the 75 percent rule applies to defining the overall category for the protected area).

To be clear, separate categorization of zones is possible when primary legislation describes and delineates zones within a protected area and not when primary legislation simply allows for zoning in a protected area, such as through a management planning process. IUCN recommends in most cases that assigning different categories to zones in protected areas is not necessary but may be relevant in larger protected areas where individual zones are themselves substantial protected areas in their own right.

Transboundary protected areas: in a growing number of cases, protected areas exist on both sides of a national or federal boundary, managed by different authorities but with some level of cooperation, varying from informal arrangements to official agreements between governments; these are known as transboundary protected areas (Sandwith *et al.* 2001). In many cases, the adjoining protected areas may be managed in different ways

and in consequence will have different categories. Whilst it is important that management approaches within the different components of a transboundary protected area are complementary, there is no reason why they should be the same.

Figure 2 outlines an example of a decision tree for deciding if a zone is suitable for having its own category.

How does ownership and management responsibility impact on the categories?

> - The category is not affected by ownership or governance

Any ownership structure or governance type can be found in any category, and examples of all combinations can be found around the world. There are some trends: large ecosystem-protection areas such as category II are more likely to be state-owned and managed while community conserved areas are probably more likely to be in the less restrictive categories V and VI, but exceptions occur. For instance some of the most strictly protected areas in the world are sacred natural sites where entrance is forbidden to all but a few specially appointed people, or in some cases no human at all is allowed to enter.

What about the areas around protected areas?

> - Buffer zones, biological corridors etc. may or may not also be protected areas (and thus eligible for a category) depending on the form of management and recognition by the state

Conservation planners stress the importance of connecting protected areas through biological corridors and stepping-stones (sympathetic habitat used by migratory species) and insulating them with buffer zones. Unfortunately competition for land, population pressure and poor governance mean that many protected areas remain as isolated "islands". Addressing this through restoration projects, compensation packages, set-asides, voluntary agreements and legislative changes is a long-term challenge. Whether or not such areas can be assigned a category depends on whether or not they qualify as protected areas under the IUCN definition. Some category V protected areas have been set up to serve as buffer zones around more strictly protected areas. Other buffer zones and biological corridors are not protected areas but are instead areas where a combination of voluntary agreements and/or compensation packages helps to protect the integrity of the protected area through landscape approaches and connectivity conservation. For example in some countries commercial tree plantations or managed natural forests help to buffer protected areas by preventing land conversion: but neither of these uses would qualify as a protected area.

Figure 2.   Zones and IUCN protected area categories

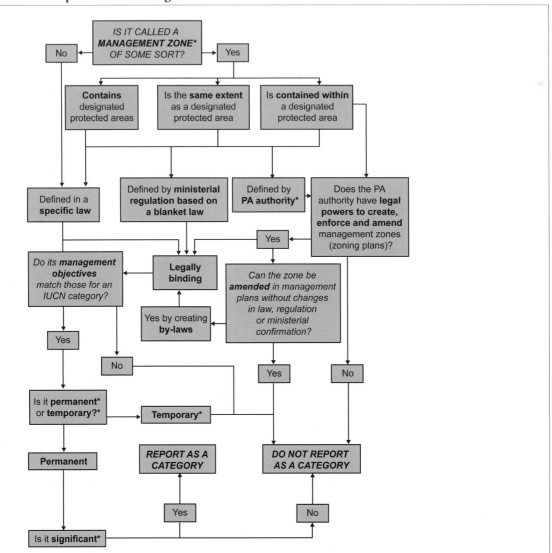

* **Management zone** – e.g., buffer zone, wilderness zone, recreation zone, no-take zone, core zone etc.
  Protected area authority – Ministerial department, agency, NGO or community institution that is recognised in law
  Permanent – inscribed in law, established and recognised, subject to a long-term vision (e.g., core zone for key breeding species)
  Temporary – established for management purposes only, temporal (e.g., for a limited period)
  Significant – of a recognisable and reasonable scale and/or proportion to the wider landscape

How do other international protection designations relate to IUCN protected areas and categories?

- Most other international protection designations are not necessarily protected areas as recognised by IUCN, although in practice many are protected areas

- World Heritage sites, Ramsar sites and Natura 2000 sites can have any or no IUCN category

- Biosphere reserves should have a highly protected core (category I–IV) and a sustainable management zone around (category V/VI or not a fully protected area)

A range of global or regional efforts exists to define conservation for areas of land and water, including:

- UNESCO World Heritage – natural and mixed natural and cultural sites agreed by the WH Committee to be of "outstanding universal value";
- UNESCO Man and the Biosphere (MAB) – biosphere reserves are sites where conservation is integrated with sustainable use;
- Ramsar sites – important freshwater and tidal waters listed by the Ramsar Convention.

Working out the relationship between these sites and IUCN protected areas is complicated and described in greater detail in a later section. For some of the above (e.g., natural World

Heritage sites) most listed sites are also protected areas. Some countries view such designations as automatically protected areas, while others do not. The general tendency seems to be that assigning full protected area status to these designations is often the best way of ensuring the long-term conservation of the site's values. This being the case, other designations can and do contain sites in all the IUCN categories: there is no particular link between a designation such as World Heritage status and any one or group of IUCN categories.

One possible exception would be the MAB biosphere reserves, which promote sustainable use around a core of highly protected land or water. In general, a biosphere reserve would have: (a) a highly protected core zone (usually category I–IV); (b) a buffer zone which might be category V or VI or, alternatively, managed land/water that would not correspond to an IUCN category; and (c) a transition zone that would not correspond to an IUCN category.

Assignment

The significance of the assignment process has increased as the categories have started to be applied as policy tools as well as ways of measurement. For instance when assignment of a particular category carries with it restrictions on land or water use under law, or dictates who can and cannot live in the area, as is the case in some countries, then the decision about which particular category applies is more significant than if they are simply being used as a statistical device. The process of assignment is up to the country or governing body concerned, but the following section outlines some principles and a proposed methodology.

Some principles for assignment

IUCN's approach to assignment of the protected area management categories is based on a series of principles, outlined below, relating to responsibility, stakeholder involvement and guarantees:

- **Responsibility:** use of the categories is voluntary and no body has the right to impose these. States usually have the final legal decision, or at least an overarching responsibility, about the uses of land and water, so it makes sense that states should decide on the protected area category as well.
- **Democracy:** nonetheless, IUCN urges states to consult with relevant stakeholders in assigning categories. Proposals are outlined below. Democratization and decentralization processes are resulting in an increasing number of sub-national governments taking responsibility for protected areas; here the local or regional government usually reports to the central government. In most private or community conserved areas, governments will often defer to the opinions of the owning and governing body regarding assignment, although some countries may have policies or laws in this regard.

- **Grievance procedure:** many stakeholders support the idea that there should be some way in which decisions about categories can be challenged. IUCN supports this, noting that final decisions about management still usually rest with the state or the landowner. Some proposals for possible grievance procedures are outlined below .
- **Data management:** information on protected areas, including the category, should be reported to the UNEP World Conservation Monitoring Centre, which coordinates the World Database on Protected Areas and compiles the *UN List of Protected Areas*.
- **Verification:** IUCN can advise on assignment and sometimes runs individual advisory missions to countries or even individual protected areas. IUCN is also considering the development of some form of verification or certification system for protected area categories, on a voluntary basis, where the managing authority wants verification that management objectives meet the assigned category.

A process for assignment

It is recommended that assignment should rest on four main elements:

- Good guidance for governments and other protected area authorities;
- An agreed process for assignment;
- A system for challenging assigned categories, to be developed;
- A process of verification; which could be implemented at the national level (by an expert panel for example) or requested from an independent body such as IUCN.

The first three are discussed below: currently a verification system does not exist although may be developed in time.

Good guidance for governments and other protected area authorities

The basis of using the categories is the guidance contained in this publication. In addition, more detailed guidance relating to specialized issues may be available or become available, for example with respect to:

- **Biomes:** e.g., forests (Dudley and Phillips 2006), marine, inland water protected areas etc;
- **Categories:** similar to the guidance developed for category V (Phillips 2002), already planned for category Ib and category VI;
- **Regions:** similar to guidance already produced in Europe (EUROPARC and IUCN 1999) and planned for several other regions, either as guidelines or case studies;
- **Selection tools:** for identifying category and governance type;
- **Governance types:** there is also interest in producing more detailed information on private protected areas, community conserved areas and indigenous peoples' protected areas.

An agreed process for assignment

Figure 3 below shows a proposed process for assignment: ideally, this should involve many stakeholders, particularly when assignment to a particular category will have impacts on people living in or near the protected area or on other stakeholders. One option would be to have a national task force reviewing data on protected areas and it has been suggested that a national committee for IUCN might be an obvious vehicle for this. The extent that stakeholders are involved in these decisions ultimately rests with governments and IUCN can only advise and encourage. A number of tools exist to identify the best category for a particular site. Sometimes questions will relate to a whole series of similar sites: for example if a forest department is trying to decide which of its forest reserves should be recognised as protected areas; or when private protected areas are trying to attain protected area recognition within national systems; or where communities are interested in converting their fishery control zones into protected areas.

Reporting

Once a category is assigned, governments are requested to report this to the UNEP World Conservation Monitoring Centre, so that information can be included in the World Database on Protected Areas (WDPA) and the *UN List of Protected Areas*. Reporting is voluntary, but is requested by a number of United Nations resolutions and policies, most recently in the CBD *Programme of Work on Protected Areas*. This implies that there are expectations on governments to report information in a regular and accurate fashion, following the template supplied by UNEP-WCMC. There are similarly obligations on UNEP-WCMC to ensure that information is transferred accurately and quickly to the database.

Figure 3.   Process for assigning protected area categories

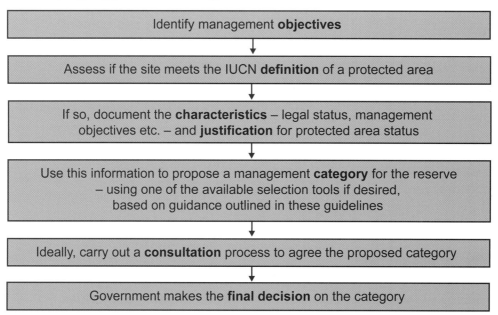

Strengthening the assignment of categories

Assignment of categories has traditionally been the responsibility of governments and it has been assumed that they and others will not wilfully assign an incorrect category and that governments have capacity to assign categories correctly. This relationship has over the last few years come under increased scrutiny and question, particularly from some industry groups that are concerned about the increasing amount of land and water "locked up" from other forms of development but also from local communities, concerned about loss of rights and access. Some governments have also asked for clarification that a particular protected area has been assigned the correct category; particularly when funding levels for protected areas have been set on the basis of category designation. It has been suggested that some kind of grievance procedure or verification process may be useful to provide an independent guarantee that: (1) the area is truly a protected area; and (2) the correct category has been assigned. Ultimately choice of category rests with individual governments and IUCN has no right or wish to impose on what should be national decisions. However, there has been strong support for IUCN to provide a framework for governments and others to strengthen and where necessary question category assignment.

One option is for IUCN, or some third party, to establish a certification or verification process aimed at checking the assignment of categories – these issues are examined in greater detail in the section on management effectiveness, mainly in terms of particular cases where verification of standards may be useful to the protected area owners or managers themselves.

A different issue relates to the possibility of external stakeholders challenging the assignment to a category. Again it is to be hoped that such instances remain rare but it is becoming clear that some system for addressing this needs to exist within IUCN and WCPA. IUCN WCPA intends to cooperate with partners, including UNEP-WCMC, to investigate practical options for implementing some kind of grievance procedure in the near future.

Such a process can only ever be symbolic: governments have the final right to say how a protected area is managed and how it is categorized. But independent assessments of this kind have proven of important political value in similar situations, such as the Ramsar *Montreux List* and the Reactive Monitoring Mechanism under the World Heritage Convention.

IUCN recognises the need to help governments and other institutions to increase their capacity in terms of understanding and applying the categories. In conjunction with the launching of the new category guidelines, IUCN is developing a major project on capacity building in their application.

5. Using the categories

The categories were originally designed as a way of classifying and recording protected areas – already a huge task. Gradually new uses have been added, including in particular a role in planning protected area systems and in developing coherent conservation policy: after initial reluctance IUCN members themselves endorsed this approach through a recommendation that governments ban mining in category I–IV protected areas.

Using the IUCN protected area categories as a tool for conservation planning

Historically the protected area management categories have been used by management agencies to classify, with varying degrees of accuracy, the purpose of a given protected area once this has been determined through conservation planning. IUCN recommends that protected area management categories also be used to help in the design of protected area systems with varying management purposes (and governance types) to meet the needs of biodiversity across the landscape or seascape. As governments are called upon to identify and fill gaps in their protected area systems, planners should apply the full suite of protected area management categories when identifying, designating, and launching management of new protected areas.

Background

As human use and consumption dominates much of the world's land and seascapes, there is a growing need to view protected areas as a *range* of management practices rather than isolated, locked-up and restricted places. A "one-size fits all" approach to the management of biodiversity in protected areas will not only create conflict with other societal needs, but will limit the management options for conservationists and the amount of land and sea available for biodiversity protection. The diversity of protected area categories can be used to tackle an ecological necessity of a species or ecosystem, and balance that with society's needs.

Under agreements of the CBD, governments are committed to completing ecologically-representative systems of protected areas, and this process usually starts by identifying **gaps** in the current system – typically through an ecological gap analysis. In a conservation context, gap analysis is a method **to identify biodiversity (*i.e., species, ecosystems and ecological processes*) not adequately conserved within a protected area system** or through other effective and long-term conservation measures. Well designed ecological gap analyses identify three types of gaps in a protected area system (Dudley and Parrish 2006):

- **Representation gaps:** no or insufficient existing coverage of a species or ecosystem by a protected area;
- **Ecological gaps:** protected area system fails to capture places or phenomena that are key to conserving a species or ecosystem during its life cycle;
- **Management gaps:** the protected areas geographically cover the biodiversity elements but fail to protect them due to insufficient or inadequate management.

When gaps are identified and resulting actions are implemented – such as new protected areas being proposed and reviews of management categories for existing protected areas being conducted – the full suite of categories should be considered.

When reviewing the categories of *existing* protected areas to determine the type of protection that will best conserve the biodiversity within that protected area, there is no hierarchy that suggests, for instance, that a category I protected area is invariably better than a category II or III or IV. On the other hand, categories are not simply interchangeable. The only principle that should apply in assigning categories is the appropriateness of a protected area's assigned management purpose within the system ***relative to the ecological needs of, and threats to, the species or ecosystem*** in the context of the entire landscape or seascape where that biodiversity occurs. The protected area objectives also need to be considered at the moment of reviewing and assigning a management category. In some cases, it may be best to increase the stringency of protection because of declines in the ecological or conservation status of a species or ecosystem within the protected area or across its distribution– e.g., part or all of a category V protected might be reassigned as a category Ib. In others, it might actually be more strategic to shift management to allow more flexibility in terms of sustainable use (e.g., from a category II protected area to a category VI).

Increasing the stringency of protection will usually be a response to a continued decline in biodiversity within an existing protected area. When might natural resource managers choose a less strictly protected area approach over a more restricted one? Examples include:

- When the viability of a species' population or the integrity of the ecosystem has improved across its distribution and no longer requires reduced human use and intense protection.
- When the potential human uses in a lower protected area category are unlikely to affect the health of the species or ecosystem.
- When changing the category increases the size of the protected area to the benefit of target species and ecosystems. For example, it may be more effective in river and freshwater protection to manage more of a watershed for ecosystem function with less restrictive protection than to protect the main stream of the river as category I or II, depending on the priority threats to the biological target.
- When biodiversity has become adapted to cultural management systems and the absence of these interventions now places pressure on species' survival or viability.

Some considerations for assigning protected area management categories in protected area system planning

There are no hard and fast rules about choosing a particular category for a given protected area. However, the over-riding approach should be to recognise that not all protected areas will be managed in the same way and that the choice of management approach needs to be made by weighing the different opportunities and pressures relating to the area. Some general principles are outlined below.

- **Start with the ecological needs of species and ecosystems.** Management options should be determined primarily by the ecological characteristics and life history of the species and ecosystems. For example, different species have different responses to disturbance and in general the most sensitive species may require stronger protection under the more restrictive management approaches.

- **Consider the threats to the species or ecosystem values.** Some threats lend themselves to a particular management approach. For example poaching in marine protected areas may be best addressed by allowing local fishing communities access to an agreed level of catch (e.g., in a category V or VI protected area) thus encouraging them to help control poaching by outsiders.

- **Consider the protected area's objectives, existing and proposed international designations and how they contribute to the landscape, country and global biodiversity conservation efforts.** Each existing protected area should have been established for specific purposes. But when the planning approach is broadened to consider the landscape and country levels it may be necessary to re-consider the original purposes and objectives. International designations such the World Heritage Convention and the Ramsar convention are useful in identifying the best approach to manage a site.

- **Consider developing and implementing a process to assign/review management categories in a country.** A national protected area agency should develop an official process to review and assign management categories. For example, as a result of an ecological gap assessment, the protected area agency in Panama reviewed the management categories of all protected areas in the country.

- **No loss of naturalness, ecosystem function, or species viability.** The management option chosen should not in most cases result in a loss of current naturalness within the protected area (e.g., IUCN would not normally propose a category V or VI protected area in a more-or-less natural site) although there may be exceptions.

- **Consider the landscape and seascape when assigning categories.** Choice of category should reflect the protected area's contribution to the overall conservation mosaic rather than just the values of the individual site, i.e., management objectives for any given site should not be selected in isolation. For example, an inland lake might not only be important for resident populations but as a staging ground for migratory birds. Similarly, we recommend that environmental planners should develop a diverse portfolio of managed areas across the IUCN categories for a given biodiversity element.

- **Stakeholders matter.** Management options should consider the needs, capacities and desires of local communities and should generally be selected after discussion with stakeholders – management objectives that are supported by local communities are more likely to succeed than those that are unpopular or opposed.

- **Consider management effectiveness when assigning protected area categories.** Managers should also take into account the existing and likely management effectiveness of a given area when recommending management purpose (protected area categories). Ineffective or non-existent management in a category I or II protected area (the paper-park syndrome) may achieve less conservation impact than an effective category V or VI protected area even if the management rules in the latter are less stringent.

- **More restrictive management categories are not always better.** Conservation scientists often assume that categories I–IV represent more effective conservation than categories V–VI in designation of protected areas. This is not always the case; for example less restrictive approaches that cover larger areas can sometimes be more effective.

- **Use the categories as a tool for within-protected area planning.** Within a single protected area, several zones with different management objectives can be agreed if this helps overall management. Consider temporary zones within protected areas (e.g., to allow low-impact sustainable exploitation of non-timber forest products by local communities).

- **Consider societal benefits of diversifying the category portfolio.** Considering a variety of protected area management categories can often improve public perceptions of protected areas and increase their likelihood of success – particularly if people recognise that not every protected area means that the terrestrial, aquatic or marine resources are "locked up". Use of certain categories can build commitment by stakeholders for conservation and expand options for designation of areas for protection (e.g., sacred sites for local people's religion that also represent significant contributions to biodiversity, as is the case in Tikal National Park, Guatemala).

Planning for climate change

Global warming will affect protected area planning in a number of ways. Climate change will bring an increase in the average annual temperature, changes in the water regime and almost certainly greater unpredictability. There are likely to be fundamental changes in the natural attributes driving ecosystems and habitats and the distribution of biotic natural features. Wetlands may dry out in some parts of the world, and elsewhere dry areas may become prone to flooding. Low-lying islands and coastal land will be more vulnerable to erosion and loss of land and habitats as a result of sea-level rise and more stormy conditions. Species and habitats at the edge of their geographical range are more likely to be adversely affected by global climate change. The seasonal rhythms of plants and animals will also change. Many protected areas are likely to be affected, potentially losing species and ecosystems; other species may come in to take their place although it is likely that many of the less mobile or adaptable species will face increased threats of extinction. But at the same time, protected areas will be able to play a role in mitigating climate change, by providing buffers against extreme climate events (Stolton *et al.* 2008) and a network of natural habitats to provide pathways for rapid migration and space for evolution and adaptation (Dudley and Stolton 2003).

Protected area managers and authorities are starting to look at the options available for reducing the impact climate change will have on protected areas and for maximizing the benefits that well-designed protected area systems can have for wider society in mitigating the impacts. In terms of management objectives and categories, this has a number of implications:

- Likely climate change impacts should be factored in when designing protected area systems to maximize the opportunities offered by a range of management approaches, based on an understanding of the strengths and weaknesses of different categories in the face of climate change. These need to be recognised in the planning of protected area systems and of individual protected areas today, to be ready for changes in the future (bearing in mind that we still often do not know with any confidence what these changes are likely to be – so planning needs to build in flexibility).
- Connecting protected areas through corridors and networks will become even more essential in order to facilitate the movement of species and increase the likelihood of natural transfers to places where conditions are more suitable for survival. Designing larger protected areas with a greater range of biogeographical characteristics will be appropriate where this is possible.
- Some species may face total extinction if there are no places within the range of their potential natural expansion where the climatic regime is suitable for their survival. It may therefore be necessary to develop schemes for the translocation of species to more appropriate locations and to improve links between *in-situ* and *ex-situ* conservation efforts.

- Climate change is likely to mean more interventionist management to protect the occurrence of species and habitats. This will raise questions about the assignment to category and perhaps greater use of category IV-type approaches.
- Changing conditions may involve alterations to management within individual protected areas. In some cases harsher conditions may render traditional cultural landscapes unsustainable and also put remaining species under threat, necessitating a change from category V to, say, category Ib, associated with abandonment by humans. In other situations, formerly fairly pristine environments may only be able to survive with human intervention and a category Ia approach might have to shift to a category IV. It should be noted that changes in category should be a rare event and subject to as rigorous a process as original assignment.
- Protected areas may need to be relocated (for example if the sea level rises) or new protected areas created; in some cases existing protected areas may become irrelevant if the species they were designed to protect can no longer survive there. We have become used to seeing protected areas as fixed entities that remain sacrosanct for the foreseeable future but under conditions of climate change this may no longer be the most effective way of implementing conservation.
- Research on climate change and protected areas should be encouraged. Such research should also assist protected area managers to develop appropriate and relevant responses to climate change.
- Wherever possible larger protected areas should be established with a greater range of biogeographical characteristics, to provide space for changes in range and buffering against extreme weather events.

Most of these strategies are beyond the scope of the current guide. However, we need to build up information about the relative merits of the different categories and how they can be used more effectively as a key element of overall response strategies to climate change. Table 5 gives a preliminary analysis of strengths, weaknesses, opportunities and threats for the categories from a climate change perspective.

Table 5. Strength-Weakness-Opportunity-Threat analysis for categories under climate change

Category	Strengths	Weaknesses	Opportunities	Threats
Category Ia	Strict protection of a pristine environment provides baseline data to measure changes and plan responses.	Often quite small, therefore with low buffering capacity.	Added stresses may need greater management intervention and a switch to e.g., a category IV approach.	Leaving a protected area completely alone may be a high risk option in the face of rapid environmental change.
Category Ib	Large areas of relatively unmodified habitat are generally thought to be strongest at absorbing changing climatic conditions – with the opportunity to protect whole ecosystems and associated processes.		A chance to maintain very large areas of unmodified habitat with minimum human intervention to allow natural adaptation to climate change.	
Category II			Space to focus on ecosystem approaches, active management already in place to facilitate this.	Many category II and III protected areas survive on tourism revenues, which may be at risk with higher fuel prices and campaigns against holiday flying.
Category III	Usually iconic sites with a high degree of commitment to continued protection.	Often too small to absorb impacts of climate change.	Can provide "islands" of protection in otherwise heavily altered landscapes.	
Category IV	Management interventions to maintain target habitats and species may already be written into site plans.	Usually fragments of habitat, likely to have relatively low resistance to changing climate.	Already human management is in place so these provide a useful laboratory to try out modifications in management.	Loss of conditions necessary for the particular species being protected.
Category V	Long-term management strategies in place.	A proportion of the habitat has already been altered and perhaps weakened (e.g., to the presence of invasive species).	Cooperation with local communities to develop adaptive management strategies in mainly cultural landscapes and seascapes.	Land abandonment due to changing conditions and therefore loss of the cultural systems on which biodiversity has come to depend. Extra pressure on resources due to harsher conditions.
Category VI	Human commitment to long-term protection.		Cooperation with local communities to develop adaptive management strategies for sustainable management.	Shifting climate renders previously sustainable management systems less viable.
All categories	Maintaining healthy ecosystems, which are judged to be the best adapted to face climate change impacts. Maintaining adaptive potential and *in-situ* gene banks.	Fixed in one location and therefore susceptible to climate shifts.	Changing management strategies in response to change, drawing on experience in other categories and in sustainable management outside protected areas.	Climate change makes the site unsuitable for target species and habitats.

Using the IUCN protected area categories as a tool for conservation policy

Although the categories were not originally intended as policy instruments, in practice they have frequently been used as such, both by IUCN itself and much more frequently by governments and other institutions. Those using the categories need to be aware of this reality and factor it into their application. There are six broad types of policy use, with varying degrees of official status:

- **International descriptive policy:** where the categories are officially adopted for recording – one of the original aims of the categories system. The categories have been adopted by the UN system, for example in the *UN List of Protected Areas*[8] and the CBD *Programme of Work on Protected Areas* and in the World Database on Protected Areas. At the international level there has also been limited use of the categories system within global institutions and agreements such as the Intergovernmental Forum on Forests, the UN Forest Resource Assessment and also within the context of biosphere reserves.
- **International prescriptive policy:** more controversially, the categories have been used in a limited way to suggest international policy including controls on particular management interventions within protected areas. Most significant was the development of an "IUCN No Go position on mining in categories I to IV". This recommendation (number 2.82) was adopted by the IUCN World Conservation Congress in Amman in 2000. It recommended, *inter alia* "*IUCN Members to prohibit by law, all exploration and extraction of mineral resources in protected areas corresponding to IUCN Protected Areas Management Categories I to IV*". This recommendation played an important role in the adoption by Shell and ICMM of a "No-Go" commitment in natural World Heritage sites. It represented a new application of the IUCN categories system in that it linked restrictions on resource use to the system itself but also raised important questions about whether the system was rigorous enough for these purposes.
- **Regional policy:** two regional conventions and agreements have applied the IUCN categories (Dillon 2004). These are the Conservation of Arctic Flora and Fauna (CAFF) Circumpolar Protected Areas Network (CPAN) Strategy and Action Plan 1996 and the Revised African Convention on the Conservation of Nature and Natural Resources 2003. In the case of the African Convention, the IUCN categories had a strong influence on the development of the revised Convention and provided a framework for a number of sections, initially being endorsed by

an interagency taskforce and then submitted to a number of African government experts, who adapted the text to the African context. Article V of the Convention defines a Conservation Area as any protected area designated and managed mainly for a range of purposes, and then goes on to elaborate these purposes by referring to the six IUCN categories. Another example of regional-level application exists within Europe, where a WCPA/EUROPARC Federation publication was prepared on *Interpretation and Application of the Protected Area Management Categories in Europe*, to provide guidance for the European context.

- **National descriptive policy:** a number of countries have made conscious efforts to align their existing protected area categorization to the IUCN system, either by changing categories to fit the system directly or by agreeing equivalents so that cross comparisons are easy. Although use of the categories is voluntary, most countries currently apply categories to some if not all of their protected areas. Some 10 percent of national protected area legislation since 1994 has used the IUCN categories. This includes legislation in Australia, Brazil, Bulgaria, Cambodia, Cuba, Georgia, Hungary, Kuwait, Mexico, Niger, Slovenia, Uruguay and Viet Nam.
- **National prescriptive policy:** a smaller subset of countries have explicitly linked policies to categories, including level of funding (e.g., Austria) or policies on settlement in protected areas. In a number of cases, countries have provided elaboration of what the categories mean in the national context, keeping to the original framework but providing policy details – as is the case in Madagascar.
- **NGO policy:** use by NGOs is less official, but nonetheless significant. For example several NGOs have in effect only "counted" categories I–IV as protected areas, thus influencing many ecoregional or bioregional plans. NGOs have also used IUCN categories for advocacy purposes, for example lobbying for particular management approaches in protected areas.

Lessons learnt from application of the categories system in policy

Experience to date has provided some general lessons about the use of the categories as policy:

- The categories have significant potential for influencing protected area policy and legislation at all levels, and the level of application has greatly accelerated since the publication of the 1994 guidelines;
- It is anticipated that the relative importance of the categories system in influencing policy decisions will increase, particularly at national levels, as the CBD *Programme of Work on Protected Areas* is more widely and effectively applied;

---

[8] The 1994 Categories were used as the basis for compiling the 1997 and 2003 versions of the UN List.

- The advantages of including the categories system in policy-level decisions are that it gives the system extra weight and credibility and can enhance awareness and understanding of the values of protected areas;
- The most effective use of the categories system in policy-level decisions has been where the system is applied in a flexible way, in response to unique national or regional circumstances;
- Application of the categories system also gives recognition in terms of international standards.

There are, however, a number of constraints to the effective application of the categories system in policy decisions. These include:

- The validity and accuracy of the process used to assign protected areas to the IUCN categories, particularly category I–IV, has been challenged: in particular related to the "no-go" policy recommendation on mining in IUCN category I–IV and suggesting that use in policy implies greater rigour in application than has been the case in the past;
- Lack of awareness and/or understanding of the IUCN categories system;
- Variable accuracy of data on protected areas in the World Database on Protected Areas and the *UN List of Protected Areas*;
- Lack of understanding and awareness of how the categories system can be applied at national levels and also in particular biomes.

It follows that future effort to use the categories in policy decisions must be based on a more rigorous understanding and objective application of these categories.

6. Specialized applications

Protected areas embrace a huge range of biomes, ownership patterns and motivations – all these impact on the way that management objectives are set and therefore on the subsequent categories that are applied. This section looks in more detail at some particular cases that have caused confusion in the past: forests, freshwater and marine protected areas, sacred natural sites and the role of restoration in protection.

Forest protected areas

There has been confusion about forest protected areas and in particular what "counts" as a protected area in the forest biome, particularly when such information is incorporated into wider data collection about forest resources. The following guidelines (based on Dudley and Phillips 2006) address a series of issues including:

- Definition of a forest in the context of forest protected areas;
- Applying the IUCN categories system to forests;
- Calculating the extent of forest protected areas;
- What areas fall outside the IUCN definition of a forest protected area?
- Distinguishing biological corridors, stepping stones and buffer zones inside.

Definition of a forest in the context of forest protected areas

The definition draws on that of UNECE/FAO and adds interpretation from IUCN as follows:

UNECE/FAO definition of forest

Forest: Land with tree crown cover (or equivalent stocking level) of more than 10 percent and area of more than 0.5 ha. The trees should be able to reach a minimum height of 5 m at maturity *in situ*. A forest may consist either of closed forest formations where trees of various storeys and undergrowth cover a high proportion of the ground, or open forest formations with a continuous vegetation cover in which tree crown cover exceeds 10 percent. Young natural stands and all plantations established for forestry purposes which have yet to reach a crown density of 10 percent or tree height of 5 m are included under forest, as are areas normally forming part of the forest area which are temporarily unstocked as a result of human intervention or natural causes but which are expected to revert to forest.

Includes: Forest nurseries and seed orchards that constitute an integral part of the forest; forest roads, cleared tracts, firebreaks and other small open areas; forest in national parks, nature reserves and other protected areas, such as those of special scientific, historical, cultural or spiritual interest; windbreaks and shelterbelts of trees with an area of more than 0.5 ha and width of more than 20 m; plantations primarily used for forestry purposes, including rubberwood plantations and cork oak stands.

Excludes: Land predominantly used for agricultural practices.

Other wooded land: Land either with a crown cover (or equivalent stocking level) of 5–10 percent of trees able to reach a height of 5 m at maturity *in situ*; or a crown cover (or equivalent stocking level) of more than 10 percent of trees not able to reach a height of 5 m at maturity *in situ* (e.g., dwarf or stunted trees); or with shrub or bush cover of more than 10 percent.

Policy guidance: The UNECE/FAO definition should be used in relation to forests in forest protected areas with the following caveats:

- Plantation forests whose principal management objective is for industrial roundwood, gum/resin or fruit should *not* be counted;
- Land being restored to natural forest *should* be counted if the principal management objective is the maintenance and protection of biodiversity and associated cultural values;
- "Cultural forests" should be included, *if* they are being protected primarily for their biodiversity and associated cultural values.

Applying the IUCN categories system to forests

Much of the potential confusion about what is or is not a protected area can be avoided if the hierarchical nature of the definition is stressed, and the system is applied sequentially. In short, the categories are only to be applied to forest protected areas if the area in question already meets the definition of a protected area. Even after a protected area has been correctly identified, mistakes are possible in deciding into which category to assign it. Two questions arise:

- **How much of a protected area should be forest before it is counted as a forest protected area?** Some important forests within protected areas may in fact be a minority habitat, such as relic forests, riverine forests and mangroves. This creates problems of interpretation and data availability. Should forest statisticians differentiate the fractions of protected areas that contain forests?
- **Is all the forest in a protected area automatically a forest protected area?** Some protected areas, particularly categories V and VI, may contain areas of trees that are not protected forests, such as the exotic plantations in many category V protected areas in Europe. **These do not meet the definition of a forest proposed for use in protected areas outlined above** but currently they are sometimes recorded as being "protected" – and thus can appear in official statistics as "forest protected areas".

It is important that a standardized procedure is followed in determining the extent of forest protected areas that gives meaningful and accurate data. Calculation should follow the sequence shown below. Forest protected areas can be calculated as an unambiguous subset of national protected area statistics, capturing information on all protected forests but eliminating plantations within the less strictly protected categories.

Policy guidance and interpretation: the process of assignment should therefore *begin* with the IUCN definition of a protected area and then be *further refined* by reference to the IUCN categories:

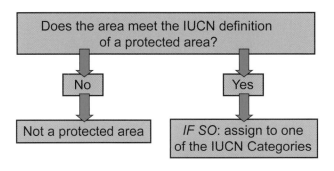

It follows that ***any area that appears to fit into one of the categories based on a consideration of its management practices alone, but which does not meet the general definition of a protected area, should*** **not** ***be considered as a protected area as defined by IUCN.***

Calculating the extent of forest protected areas

When statistics are required that relate specifically to forests, it is necessary to identify that portion of protected areas that contains forest. This will seldom be straightforward: many protected areas contain some forest, even "forest protected areas" are often not entirely forest and in addition calculation sometimes needs to take into account forests within broader-scale landscape protection that do not meet the identification criteria listed above.

Policy guidance: calculation of forest protected area includes the following steps:

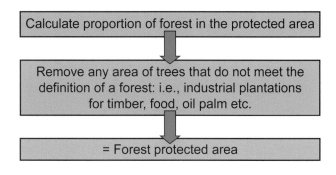

What areas fall outside the IUCN definition of a forest protected area?

There are many forest uses – some with high social and ecological or biological values – that lie outside the IUCN definition.

Policy guidance: the following are *not* automatically forest protected areas:

- Forests managed for resource protection other than biodiversity – e.g., forests set aside for watershed or drinking water protection, avalanche control, firebreaks, windbreaks and erosion control;
- Forests managed primarily as a community resource – e.g., forests managed for non-timber forest products, fuelwood and fodder, for recreational or religious purposes;
- Forests managed as a strategic resource – e.g., as an emergency supply of timber in times of conflict;
- Forests with unclear primary management objectives resulting in biodiversity protection being considered as an equal or a lesser priority along with other uses;
- Forests set aside by accident – e.g., woodland in the central reservation or verges of motorways, forest maintained for military or security reasons.

Some examples are given in Table 6.

Table 6.   Examples of Forest Protected Areas, and also of well conserved forests that are not Forest Protected Areas

Type of forest	Example	Notes
Examples of Forest Protected Areas		
IUCN category Ia protected area	Wolong Nature Reserve, Sichuan, China	A strict protected area, established primarily to protect the giant panda, including a captive breeding centre.
IUCN category II protected area	Huerquehue National Park, Chile	This national park is entirely protected (there are some properties within it, but excluded from the protected area, that are used for ecotourism). It was established mainly for the preservation of the unique *Araucaria* (monkey puzzle) forests.
IUCN category III	Monterrico Multiple Use Area, Guatemala	This is a coastal area with the largest remaining block of mangrove in the country, plus turtle beaches and several marine communities. Mangroves are managed for protection and artisanal fishing.
IUCN category IV	Dja Faunal Reserve, Cameroon	In the southeast of Cameroon in the Congo Basin. Many people live in and around the protected area including tribes of *baka* (pygmy) people. Active management is needed to control the bushmeat trade and to help restore areas of forest.
IUCN category V	Sugarloaf Mountain, Brecon Beacons National Park, UK	The woods on the side of the mountain are owned and managed as a protected area by the National Trust, a large UK NGO, although limited sheep grazing is permitted within the forest protected area. Surrounding hills are used for sheep pasture.
IUCN category VI	Talamanca Cabécar Anthropological Reserve, Costa Rica	Some forest use is permitted in this protected area, particularly by indigenous peoples, but most of it remains under strict protection.
Examples of forests that are not Forest Protected Areas		
Forest in IUCN category V	Plantation forest within the Snowdonia National Park, Wales, UK	Although the plantation is within the category V protected area, it is an entirely commercial, state-owned timber plantation of exotic species and as such does not constitute a forest protected area.
Forest managed for environmental control	Brisbane watershed, Queensland, Australia	Some parts of the catchment around Brisbane are set aside from logging and other disturbance so as to maintain the city's water supply. The forest is strictly conserved but not as a protected area as there is no special purpose of biodiversity protection, although there are protected areas that make up a component of the catchment as well.
Forest managed by the community .	The local community in Kribi, south-west Cameroon	Local people are managing a forest under a project being facilitated by WWF. The forest seeks to provide benefits to both local people and the environment, but is not designated as a protected area (and does not have special biodiversity protection aims).
Forest managed for multiple purposes	Forests of the Jura Mountains, Switzerland	Swiss forest policy stresses multiple purpose management, selective logging and conservation. The Jura is a valuable resource for both local communities and wildlife. However, the region as a whole is not a protected area, although there are some protected areas (of various categories) within it.
Forests protected by accident	Forests on the border between South and North Korea (the de-militarized zone)	Large areas of forest are completely conserved by exclusion for defence purposes, but this situation could alter if there is a political change.

The Seychelles remote Category Ia Aldabra Atoll in the Western Indian Ocean provides an ideal natural laboratory for studying tropical marine ecosystems and related environments (such as seagrass and mangroves). © *Sue Stolton*

The Imfolozi Wilderness Area (in the Imfolozi Game Reserve, KwaZulu Natal, South Africa) is a provincially managed category 1b area from which came the impetus to create other wilderness areas in Africa. Here, "trailists" with the Wilderness Leadership School visit the area on a five-day walking trail that utilizes low-impact camping practices. © *Vance G. Martin*

Plate 1

Kaziranga is a classic category II National Park. Famous for the Great Indian one-horned rhinoceros, the landscape of Kaziranga can be enjoyed by tourists on elephant rides or boat trips on the Brahmaputra River. © *Nigel Dudley*

Organ Pipe Cactus National Monument (category III) protects the majority of the organ pipe cactus found in the United States. © *Nigel Dudley*

Plate 2

Covering less than 1 km², the category IV Insel Vilm Nature Reserve has some of the oldest oak and beech woods in Germany; visitation is strictly controlled and much of the island is closed to human presence. © *Sue Stolton*

The category V Snowdonia National Park in Wales protects extensive areas of windswept uplands and jagged peaks within a cultural landscape, dominated by the impacts of pastoralism and the former mining industry. © *Nigel Dudley*

Plate 3

The Mamirauá Sustainable Development Reserve (category VI) in Brazil is part of a large conservation complex (over 6 million hectares) in the Amazon Basin. Its management balances the need to conserve biodiversity whilst providing options to enhance the sustainable livelihoods of local people. © *Jim Barborak*

The Category II Grampians National Park in Victoria, Australia protects 975 vascular species; one third of the State's flora, 148 species of which are threatened in Victoria. © *Nigel Dudley*

Plate 4

The highly productive waters protected by the Atol das Rocas Nature Reserve (Category Ia, Brazil) provide feeding grounds for species such as tuna, billfish, cetaceans, sharks and marine turtles as they migrate to the Eastern Atlantic coast of Africa. © *Pedro Rosabal*

The only remaining rainforest areas in Singapore are protected in the Bukit Timah Nature Reserve (164 ha) and the adjacent Central Catchment Nature Reserve (about 2000 ha), both category IV protected areas. Together they comprise less than 4 percent of the original rainforest. © *Nigel Dudley*

Plate 5

The Kogelberg Biosphere Reserve was the first Biosphere Reserve to be declared in southern Africa and forms part of UNESCO's worldwide network of Biosphere Reserves. The reserve boasts 1300 different plant species in 10,000 km² – the highest plant diversity in the world.
© *Nigel Dudley*

Reserve de Geumbeul is a small community conserved area in Senegal protecting coastal mangroves, breeding populations of a giant tortoise and the southern oryx (*Oryx gazella*). © *Nigel Dudley*

Plate 6

Discussing zoning of protected areas in Catalonia, Spain. A network of protected areas in different categories helps to maintain the biodiversity of this rich Mediterranean landscape. © *Nigel Dudley*

Yellowstone National Park (category II) in the USA is a landscape continually being shaped by geological forces. Yellowstone holds the planet's most diverse and intact collection of geysers, hot springs, mudpots and fumaroles. © *Roger Crofts*

Plate 7

It has been estimated that in southern Africa (Botswana, Namibia, South Africa and Zimbabwe) there is some 14 million ha of private land under some form of wildlife protection or sustainable wildlife management. © *Nigel Dudley*

Nyika National Park (category II) in Malawi contains several sacred natural sites and also important remnant rock art as well as high levels of native biodiversity. © *Nigel Dudley*

Plate 8

Distinguishing biological corridors, stepping-stones and buffer zones inside and outside forest protected areas

IUCN also suggests guidelines for identifying when some important linking habitats – such as corridors and buffer zones – fall inside or outside definitions of a protected area (see Table 7 below).

Table 7. Distinguishing connectivity conservation areas such as biological corridors, stepping-stones and buffer zones inside and outside protected areas

Element	Description	Examples
Biological corridor	Area of suitable habitat, or habitat undergoing restoration, linking two or more protected areas (or linking important habitat that is not protected) to allow interchange of species, migration, gene exchange etc.	**Protected areas** ● Designation of a forest linking two existing protected forests as a fully protected area with an IUCN category **Not protected areas** ● Areas of forest certified for good management between forest protected areas ● Area of woodland connecting two protected areas voluntarily managed for wildlife by landowner on a temporary basis ● Areas of forest covered by a conservation easement held by government or private conservation organization
Ecological stepping-stone	Area of suitable habitat or habitat undergoing restoration between two protected areas or other important habitat types that provides temporary habitat for migratory birds and other species.	**Protected areas** ● Relic forests managed to provide stopping-off points for migrating birds **Not protected areas** ● Woodlands set aside by farmers under voluntary agreements and government compensation to provide temporary habitat for migrating birds
Buffer zone	Area around a core protected area that is managed to help maintain protected area values.	**Protected area** ● Forest at the edge of a protected area that is open to community use under nature-friendly controls that do not impact on the aim of conservation. Typically a category V or VI protected area surrounding a more strictly protected core (I–IV). In some countries, buffer zones are legally declared as part of the protected area. **Not a protected area** ● Forest area outside a protected area that is managed sensitively through agreements with local communities, with or without compensation payments.

Marine protected areas

Marine protected areas (MPAs) by their nature present a particular suite of management challenges that may need different approaches to protected areas in terrestrial environments. Some of the particular characteristics of protected areas in the marine realm, which are often absent or relatively uncommon on land, are that:

● MPAs are designated in a fluid three-dimensional environment; in some instances, different management approaches may be considered at different depths (see discussion in point 3 below);
● There are usually multidirectional flows (e.g., tides, currents);
● Tenure is rarely applicable in the marine environment; more often than not, marine areas are considered to be

"the commons" to which all users have a right to both use and access;
● Full protection may only be necessary at certain times of the year, for example to protect breeding sites for fish or marine mammals;
● Controlling entry to, and activities in, MPAs is frequently particularly difficult (and often impossible) to regulate or enforce, and boundaries or restrictions over external influences can rarely be applied;
● MPAs are subject to the surrounding and particularly "down-current" influences, which often occur outside the area of management control and it is even more difficult to manage marine areas as separate units than it is on land;
● The scales over which marine connectivity occurs can be very large.

Today there are around 5,000 MPAs and many have been assigned to one or more IUCN categories. However application of the categories in the marine environment is currently often inaccurate. In addition, in situations where protected areas cover both land and sea, marine objectives are often not considered when assigning the site's category. Such inconsistencies between similar MPA types reduce the efficacy and relevance of the system as a global classification scheme. This section of the guidelines is intended to help increase accuracy of assignment and reporting.

General principles for applying categories to MPAs (or a zone within a MPA)

1. Distinguishing MPAs from other areas that are managed for some form of conservation

For an area to be regarded as a marine protected area, it needs to meet the overall IUCN definition of a protected area; some sites that are set aside primarily for other purposes (e.g., for defence purposes) may have value for marine biodiversity but would not be classified as marine protected areas.

This definition of a MPA used by IUCN since 1999 has been: "*Any area of intertidal or subtidal terrain, together with its overlying water and associated flora, fauna, historical and cultural features, which has been reserved by law or other effective means to protect part or all of the enclosed environment*" (Kelleher 1999).

The new overall IUCN protected area definition (see page 8) now supersedes the 1999 MPA definition in marine areas. Although it loses the specific reference to the marine environment, it does ensure a clearer demarcation between conservation-focused sites and those where the primary purpose is extractive uses i.e., fisheries management areas. It does not preclude the inclusion of relevant fishery protection zones but they need to be consistent with the new definition to be included as an MPA by IUCN/WCPA-Marine. Thus all areas of the sea that are dedicated in some way to conservation will qualify and for those that do not, there is clarity on how to move forward to achieve formal recognition by IUCN as a MPA.

As with terrestrial protected areas, a wide range of governance types exists. For example, many small community-managed MPAs have been set up particularly in the Pacific and SE Asia. These currently are not always recognised as MPAs by the national agencies and thus may not feature on national or international lists, or be allocated categories. One example is Western Samoa, where a network of over 50 small village fish reserves has been established under the Village Fisheries Management Plan (Sulu *et al.* 2002). The IUCN categories are intended to apply to any kind of legal or other effective management approach, and community-managed marine protected areas can be recognised as protected areas and categorized according to their management objectives provided they meet the protected area definition.

2. Temporary protection

Some sites, such as fish spawning aggregation areas or pelagic migratory routes, are critically important and the species concerned are extremely vulnerable at specific and predictable times of the year, while for the rest of the year they do not need any greater management than surrounding areas. The Irish Sea Cod Box, for example, is designed to conserve cod stocks in the Irish Sea by restricting fishing activities during the spawning period. The EU has encouraged the establishment of such conservation "boxes" within which seasonal, full-time, temporary or permanent controls are placed on fishing methods and/or access. These would qualify as MPAs if they meet the protected area definition.

3. Application of categories in vertically-zoned MPAs

In a three-dimensional marine environment, a few jurisdictions have introduced vertical zoning (e.g., different rules within the water column than those allowed to occur on the seafloor) which will result in different IUCN categories at different depths in the water column. While this may be one way of aiming for increased benthic protection while allowing pelagic fishing, it does create challenges for enforcement purposes, and vertical zonation is not easily shown within the existing two-dimensional databases or on maps. More importantly, the linkages between benthic and pelagic systems and species may not be well known, so the exploitation of the surface or mid-water fisheries may have unknown ecological impacts on the underlying benthic communities. WCPA-Marine discourages three-dimensional zoning for these reasons. For the handful of MPAs where this situation occurs, IUCN's current advice is that the MPAs should be categorized according to the least restrictive of the management regimes. For example, if the benthic system is strictly protected and the pelagic area is open to managed resource use compatible with category VI, the whole area should be assigned a category VI. This does underplay the higher level of protection given (and obscures the original benthic protection objective). However, only a handful of sites are affected in this way and use of the least restrictive category probably reflects the ecological uncertainty of whether higher levels of benthic protection are effective in these circumstances.

4. The use of zoning in multiple-use MPAs

MPAs typically comprise fluid and dynamic marine ecosystems, have a high diversity of habitats and species within an area and contain highly migratory marine species. This complexity often dictates the need for multiple objectives and complex management schemes. In the marine environment, this is particularly important and zoning is recommended in the IUCN best practice guidelines on MPAs as the best way of managing multiple-use marine areas (Kelleher 1999; Day 2002).

Multiple-use MPAs may have a spectrum of zones within them, each zone type having different objectives with some allowing greater use and removal of resources than others (e.g., no-take zones are commonly designated as one of the zones of a multiple-use MPA).

WCPA has recognised the problem of handling zones in the categories system. As in terrestrial protected areas, single management units in MPAs can be separately reported on, and accounted for, if:

- the areas concerned were defined in the primary legislation or a legislated management plan;
- these areas are clearly defined and mapped;
- the management aims for the individual zones are unambiguous, allowing assignment to a particular protected area category.

It is proposed that this approach should only be used for large, multiple-use MPAs where the zones are legally defined and make up more than 25 percent of the total area (see page 35 for an explanation of the "75 percent rule").

The identification of zones in MPAs should be based on the best available science and judgement, and also should be developed following consultation with relevant stakeholders.

By way of example, the amended entry for the Great Barrier Reef in the *UN List of Protected Areas* produced by UNEP-WCMC is proposed as shown in Table 8:

Table 8.   Categorization of the Great Barrier Reef

Area	IUCN category	Size (ha)
Great Barrier Reef Marine Park comprising:		34,440,000
Great Barrier Reef	Ia	86,500
Great Barrier Reef	II	11,453,000
Great Barrier Reef	IV	1,504,000
Great Barrier Reef	VI	21,378,000
Commonwealth Islands[9]		18,500

5. Applying different categories in MPAs

Any of the categories can be applied in marine environments, although some may be more suitable than others. Table 9, whilst not definitive, gives some indications of the range of management approaches and where they might be applied. This supplementary guidance should be read in conjunction with the broader descriptions for each category in these guidelines.

Table 9.   Application of categories in marine protected areas

Category	Notes relating to use in MPAs
Ia	The objective in these MPAs is preservation of the biodiversity and other values in a strictly protected area. No-take areas/marine reserves are the specific type of MPA that achieves this outcome. They have become an important tool for both marine biodiversity protection and fisheries management (Palumbi 2001; Roberts and Hawkins 2000). They may comprise a whole MPA or frequently be a separate zone within a multiple-use MPA. *Any* removal of marine species and modification, extraction or collection of marine resources (e.g., through fishing, harvesting, dredging, mining or drilling) is not compatible with this category, with exceptions such as scientific research. Human visitation is limited, to ensure preservation of the conservation values. Setting aside strictly protected areas in the marine environment is of fundamental importance, particularly to protect fish breeding and spawning areas and to provide scientific baseline areas that are as undisturbed as possible. However such areas are extremely difficult to delineate (the use of buoys can act as fish-aggregating devices, nullifying the value of the area as undisturbed) and hence difficult to enforce. Whenever considering possible category Ia areas, the uses of the surrounding waters and particularly "up-current" influences and aspects of marine connectivity, should be part of the assessment criteria. Category Ia areas should usually be seen as "cores" surrounded by other suitably protected areas (i.e., the area surrounding the category Ia area should also be protected in such a way that complements and ensures the protection of the biodiversity of the core category Ia area).
Ib	Category Ib areas in the marine environment should be sites of relatively undisturbed seascape, significantly free of human disturbance, works or facilities and capable of remaining so through effective management. The issue of "wilderness" in the marine environment is less clear than for terrestrial protected areas. Provided such areas are relatively undisturbed and free from human influences, such qualities as "solitude", "quiet appreciation" or "experiencing natural areas that retain wilderness qualities" can be readily achieved by diving beneath the surface. The issue of motorized access is not such a critical factor as in terrestrial wilderness areas given the huge expanse of oceans and the fact that many such areas would not otherwise be accessible; more important, however, is minimizing the density of use to ensure the "wilderness feeling" is maintained in areas considered appropriate for category Ib designation. For example, fixed mooring points may be one way to manage density and limit seabed impacts whilst providing access.

---

[9]   Note the Commonwealth Islands are legally part of the GBR Marine Park, whereas most other islands, that are under State jurisdiction, are not.

Table 9.   Application of categories in marine protected areas (cont.)

Category	Notes relating to use in MPAs
II	Category II areas present a particular challenge in the marine environment, as they are managed for "ecosystem protection", with provision for visitation, recreational activities and nature tourism. In marine environments, extractive use (of living or dead material) as a key activity is generally not consistent with the objectives of category II areas. This is because many human activities even undertaken at low levels (such as fishing) are now recognised as causing ecological draw-down on resources, and are therefore now seen as incompatible with effective ecosystem protection. Where such uses cannot be actively managed in a category II area to ensure the overall objectives of ecosystem protection are met, consideration may need to be given to whether any take should be permitted at all, or whether the objectives for the reserve, or zone within the reserve, more realistically align with another category (e.g., category V or VI) and should be changed. The conservation of nature in category II areas in the marine environment should be achievable through protection and not require substantial active management or habitat manipulation.
III	The protection of natural monuments or features within marine environments can serve a variety of aims. Localized protection of features such as seamounts has an important conservation value, while other marine features may have cultural or recreational value to particular groups, including flooded historical/archaeological landscapes. Category III is likely to be a relatively uncommon designation in marine ecosystems.
IV	Category IV areas in marine environments should play an important role in the protection of nature and the survival of species (incorporating, as appropriate, breeding areas, spawning areas, feeding/foraging areas) or other features essential to the well-being of nationally or locally important flora, or to resident or migratory fauna. Category IV is aimed at protection of particular species or habitats, often with active management intervention (e.g., protection of key benthic habitats from trawling or dredging). Protection regimes aimed at particular species or groups, where other activities are not curtailed, would often be classified as category IV, e.g., whale sanctuaries. Time-limited protection, as in the case of seasonal fishing bans or protection of turtle nesting beaches during the breeding season, might also qualify as category IV. Unlike on land where category IV may include fragments of ecosystems, in the marine environment, use of this category has a significant opportunity for broader-scale ecosystem protection, most frequently encompassing patches of category Ia or b and category II interest.
V	The interpretation of the seascape concept in protected areas is attracting increasing interest. Category V protected areas stress the importance of the "interaction of people and nature over time" and in a marine situation, Category V might most typically be expected to occur in coastal areas. The preservation of long-term and sustainable local fishing practices or sustainable coral reef harvesting, perhaps in the presence of culturally-modified coastal habitats (e.g., through planting coconut palms) could be a suitable management mosaic to qualify as category V.
VI	MPAs that maintain predominantly natural habitats but allow the sustainable collection of particular elements, such as particular food species or small amounts of coral or shells for the tourist trade, could be identified as category VI. The point where an area managed for resource extraction becomes a category VI marine protected area may sometimes be hard to judge and will be determined ultimately by reference to whether the area meets the overall definition of a protected area or not, as well as whether the area achieves verifiable ecological sustainability as measured by appropriate metrics.

The extent of extractive activities and the level to which they are regulated is an important consideration when determining the appropriate IUCN category to an MPA (or zone within an MPA). Extractive use including any type of fishing is not consistent with the objectives of categories Ia and Ib, and unlikely to be consistent with category II.

6. Classifying MPAs by what they do and not by the title of the category

Assignment of a MPA to an IUCN category should be based on consideration of management objectives, rather than the names of the categories. The same name or title for a MPA may mean different things in different countries. For example, the term "sanctuary", as used in the United States context, is a multiple-use MPA that is designated under the National Marine Sanctuary Program (e.g., Florida Keys National Marine Sanctuary). However "sanctuary" takes on a very different meaning elsewhere – in the UK, the term has been used to refer to strictly protected marine reserves in which

all extractive use is prohibited. As with terrestrial and inland water protected areas, categories are independent of names in MPAs.

Inland water protected areas

Inland water ecosystems occupy only a small area of the planet but are perhaps the most heavily impacted and threatened by human activities of all biomes and habitats. Governments and the conservation community have made commitments to conserve inland water species and habitats equal to those for the marine and terrestrial realms, but those commitments have yet to be fully realized. Moreover, in conserving these quality habitats, a critical service is being provided to people who are facing increasing shortages of potable/useful water. Inland water considerations therefore need to be integrated into the management of *all* relevant protected areas, which themselves need to be managed with respect to their wider bioregional and catchment context.

Definitions: Inland wetlands, freshwater systems, and wetlands

The terms *inland waters (inland wetlands)*, *freshwater systems*, and simply *wetlands* are often used interchangeably, but there are some differences. *Inland waters or inland wetlands* refers to all non-marine aquatic systems, including inland saline and brackish-water systems; whether transitional systems like estuaries are included is a matter of interpretation. *Inland wetlands* is the term used by the CBD. *Freshwater* is technically defined as "of, relating to, living in, or consisting of water that is not saline". Technically, then, it excludes inland saline and brackish-water systems, but in practice the term is often used as equivalent to inland wetlands. The Ramsar Convention defines *wetlands* as "*areas of marsh, fen, peatland or water, whether natural or artificial, permanent or temporary, with water that is static or flowing, fresh, brackish or salt, including areas of marine water the depth of which at low tide does not exceed six metres*". In some regions of the world the term *wetlands* is informally understood to exclude non-vegetated aquatic systems like streams, lakes and ground waters. For the purposes of these guidelines we use the term *inland waters* to describe the variety of aquatic and semi-aquatic habitats, and their associated species, that fall outside marine classifications. Natural inland water wetlands include (modified from the *Millennium Ecosystem Assessment*, Wetlands and Water Synthesis Report, Table 3.1):

- Permanent and temporary rivers and streams;
- Permanent lakes;
- Seasonal lakes, marshes, and swamps, including floodplains;
- Forested wetlands, marshes, and swamps, including floodplains;
- Alpine and tundra wetlands;
- Springs, oases and geothermal wetlands;
- Underground wetlands, including caves and groundwater systems.

Complexities of inland water protection

The relationship between protected areas and inland water conservation is complex. There are many real and perceived incompatibilities and challenges that arise when considering this relationship, including:

- **Landscape relationship and role.** Inland water systems are part of the larger terrestrial landscape and distinct parts are linked to their upstream catchments[10] through a variety of above- and below-ground hydrological processes. The prospect of "fencing off" wetland systems is in most cases technically infeasible, for the reasons described below. The most effective protected areas for inland water conservation will be part of integrated river basin management (IRBM), sometimes called integrated catchment or watershed management. IRBM involves a landscape-scale strategy to achieve environmental, economic and social objectives concurrently. IRBM is a form of the *Ecosystem Approach*, which the State Parties to the CBD have committed to implement. The world's governments are also committed to planning and implementing integrated water resources management (IWRM), which is similar in theory to IRBM but not geographically bound by river basins. In practice, regrettably, IWRM and even IRBM have not always given adequate attention to inland water biodiversity conservation.

- **Hydrological processes.** The "key driver" in running-water (lotic) inland water systems is the flow[11] regime: the magnitude, frequency, timing, duration, and rate of change of water flows. In standing-water (lentic) systems, the master variable is typically the hydroperiod: the seasonal and cyclical pattern of water. Both flow regime and hydroperiod characterize a system's "hydropattern". For nearly all inland water systems, water is generated "outside" the systems themselves and enters via overland and sub-surface pathways and tributary inflows. Protecting the hydropattern requires protection or management that extends upstream and upslope and often even into groundwatersheds.[12] In many cases, transboundary water management may be required, even if the protected area in question sits only in one state. In the case of most existing protected areas, this translates into working with stakeholders and partners to manage flow regimes outside protected area boundaries.

- **Longitudinal connectivity.** Streams and stream networks have a linear, or longitudinal, dimension along with lateral, vertical and temporal dimensions. Protecting longitudinal connectivity – the linkages of habitats, species, communities, and ecological processes between upstream and downstream portions of a stream corridor or network – is often an essential goal of inland water conservation and involves preventing or removing physical and chemical barriers. Protecting longitudinal connectivity is also identified as critical to maintaining resilient systems in the face of climate change. Conversely, additional artificial connectivity, as occurs in inter-basin transfers, can be deleterious because of alien species invasions. Traditional protected areas are often envisioned as polygons rather than linear features and are rarely designed around protection and management of the

---

[10]   A *catchment* is defined here as all lands enclosed by a continuous hydrologic-surface drainage divide and lying upslope from a specified point on a stream; or, in the case of closed-basin systems, all lands draining to a lake.

[11]   *Flow* is defined here as the volume of water passing a given point per unit of time.

[12]   The underground equivalent of a watershed, or surface water catchment.

longitudinal connectivity of stream channels. Often, stream channels are used to demarcate the boundaries of protected areas, without receiving dedicated protection themselves.

- **Lateral connectivity.** The lateral connections between streams and the surrounding landscape are essential to the ecological health of both the streams and the associated floodplain and riparian communities. These connections are driven in large part by the hydrological processes described above; with the interaction between stream flows and riparian lands creating the dynamic conditions that are the basis for the unique and rich habitats of floodplains and riparian wetlands. These lands also contribute critical organic and inorganic materials to streams, and can buffer aquatic habitats from pollutants. The width of these areas varies greatly, from relatively narrow strips in areas of steep slopes to extremely large floodplains. Protected areas can play an important role in conserving riparian and floodplain habitats and their connectivity with river channels.

- **Groundwater-surface water interactions.** Protecting above-ground inland water species and habitats usually requires looking beyond surface hydrology. Groundwater-fed systems are common in many areas, requiring protection of groundwater flows as well as surface waters. Most surface waters also depend on groundwaters (the water table) for their functioning, irrespective of whether fed by groundwater or not. Groundwaters, such as in karstic areas, provide habitat for often-specialized species as well as water for millions of people. Groundwatersheds and surface water catchments may not spatially or geopolitically coincide, adding an additional layer of complexity to protecting inflows.

- **Exogenous threats.** Inland waters generally sit at the lowest points on the landscape and consequently receive disturbances that are propagated across catchments and transmitted through water (e.g., pollution, soil erosion and eutrophication). While all protected areas must contend with threats originating outside their boundaries, those conserving inland water systems must explicitly address upslope, upstream and, in some cases, even downstream threats (such as invasive species).

- **Exclusion from inland water resources.** Human communities have always settled in proximity to inland water systems, which provide a wide array of essential ecosystem services. The fundamental right of access to fresh water, both within and upstream of protected areas, can be in conflict with the aims of some protected area categories that limit human resource use.

- **Multiple management authorities.** In many if not most countries there are overlapping and potentially conflicting responsibilities of different government agencies as they relate to the management of freshwater resources, wetland species, aquatic habitats, surrounding landscapes, and

protected areas. Consequently, managing inland water species and habitats within a protected area – which as noted above will likely require managing lands and water outside the protected area as well – can be complicated by the need to coordinate activities between multiple authorities, some with mandates at odds with biodiversity conservation.

In short, challenges abound. While, ideally, protected areas established to conserve inland water ecosystems will encompass entire catchments, more typically innovative combinations of protected areas and other strategies will need to be applied within an IRBM framework. Existing protected areas designated and designed to protect terrestrial ecosystems no doubt confer some benefits to wetland biodiversity through landscape management, but there are significant opportunities to provide enhanced protection. Designs for new protected areas can and should include inland water considerations from the outset to achieve better integration. The following pages provide introductory guidelines for how the range of different management approaches in protected areas represented by the categories can better assist inland wetland conservation.

Applying the new PA definition

The new PA definition – *A clearly defined geographical space, recognised, dedicated and managed, through legal or other effective means, to achieve the long-term conservation of nature with associated ecosystem services and cultural values* – is more inclusive of fresh waters than the previously adopted definition through its replacement of "*area of land and/or sea*" with "*a clearly defined geographic space*". Protected areas that primarily conserve inland water features such as river corridors or lakes are now clearly covered by the definition. This includes some types of protected areas that are unique to inland water ecosystems, such as designated free-flowing rivers.[13] A wide range of inland water conservation strategies targeted at protecting water quality and quantity, such as managing for environmental flows[14] and applying wise management practices to land use, normally fall outside the protected area definition. They are mentioned here because effective conservation of inland water systems within protected areas will in most cases only be achieved through coordinated use of such strategies beyond protected area boundaries.

Applying PA categories

Any of the categories can in principle apply to areas with explicit inland wetland conservation objectives. Examples of protected areas that have clear objectives relating to inland wetland conservation are found within every IUCN category (Table 10):

---

[13] Wild and scenic rivers are covered under separate legislation in some countries.

[14] The quality, quantity and timing of water flows required to maintain the components, functions, processes and resilience of aquatic ecosystems which provide goods and services to people.

Table 10.   **Examples of protected areas in different categories providing benefits to inland waters**

Category	Example	Description
Ia	Srebarna Nature Reserve (Bulgaria)	A 600 ha biosphere reserve, World Heritage site (WHS), and Ramsar site to protect Srebarna Lake, on the Danube floodplain. The reserve was set up primarily to protect the rich avifauna, especially waterfowl.
Ib	Avon Wilderness Park (Australia)	A 39,650 ha wilderness park covering entire catchments of the Avon River headwaters, designated for conservation and self-reliant recreation under the National Parks Act.
II	Pantanal National Park (Brazil)	A 135,000 ha national park (and Ramsar site) situated in a large depression functioning as an inland delta. The area consists of a vast region of seasonally flooded savannahs, islands of xerophytic scrub, and humid deciduous forest.
III	Ganga Lake (Mongolia)	A 32,860 ha natural feature (and Ramsar site) encompassing a small brackish lake and associated lakes in eastern Mongolia within a unique landscape combining wetlands, steppe and sand dunes. The lake district is of great importance for breeding and stop-over water birds.
IV	Koshi Tappu (Nepal)	A 17,500 ha wildlife reserve running along the Sapta Kosi River and consisting of extensive mudflats and fringing marshes. The reserve contains Nepal's last surviving population of wild water buffalo.
V	Big South Fork (USA)	This national river and recreation area encompasses 50,585 ha of the Cumberland Plateau and protects the free-flowing Big South Fork of the Cumberland River and its tributaries. The area has largely been protected for recreational opportunities.
VI	Titicaca (Peru)	A 36,180 ha national reserve established to protect the world's highest navigable lake.

Inland waters may be zoned to permit different levels of use. For example, in Lake Malawi National Park (Malawi), traditional fishing methods aimed at catching migratory fish are permitted in limited areas, while in most of the park the resident fish may not be fished.

Whether and how protected area categories are linked to place-based protections is case-specific. Table 11 lists a number of place-based strategies and identifies when they are particularly compatible, not incompatible, or incompatible with IUCN protected area categories. These assignments are generalities, and exceptions will exist. World Heritage sites, Ramsar sites, and biosphere reserves are included because they have been used widely to protect inland water features and because they have made zoning a management tool.

Table 11.   Compatibility of various inland water protection strategies with IUCN categories

Legend for compatibility cells: ■ = Particularly compatible with the protected area category; ▨ = Not incompatible with the protected area category; □ = Not particularly or never suitable for the protected area category.

Type of protected area: descriptions normally refer to these types as isolated entities – all can be incorporated as part of larger reserves	Compatibility with protected area category							If occurring outside I–VI, likelihood of contribution to conservation in IRBM*	Examples
	Ia	Ib	II	III	IV	V	VI		
Designation/recognition under an international convention or programme									
World Heritage site	■	■	■	■	▨	▨	□	Low	Lake Malawi (Malawi)
Ramsar site	■	■	■	■	■	■	■	Very high	Upper Navua Conservation Area (Fiji)
Biosphere reserve	■	■	■	■	■	■	■	High	Dalai Lake (China)
Freshwater place-based protection mechanisms									
Free-flowing river	▨	■	■	■	■	▨	□	High	Upper Delaware River (USA)
Riparian reserve/buffer	□	□	▨	□	■	■	■	High	Douglas River/Daly River Esplanade Conservation Area (Australia)
Floodplain reserve	■	▨	▨	■	▨	□	■	High	Pacaya-Samiria (Peru)
Fishery/harvest reserve	□	□	□	□	▨	▨	■	High	Lubuk Sahab (Indonesia)
Wetland game/hunting reserve	□	□	□	□	■	■	■	Moderate	Ndumo Game Reserve (South Africa)
Recreational fishing restricted area	□	□	□	□	□	□	■	Moderate	Onon River (Mongolia)
Protected water supply catchment	■	■	■	□	▨	■	■	High	Rwenzori Mountains National Park (Uganda)
Protected aquifer recharge area	□	□	□	□	■	□	□	High	Susupe Wetland (Saipan)
Other place-based mechanisms with potential freshwater benefits									
Marine reserve/coastal management zone	■	▨	■	■	■	□	□	Low	Danube Delta (Romania)
Seasonally closed fishery	□	□	□	□	▨	▨	■	Moderate	Lake Santo Antonio (Brazil)
Forest reserve	□	□	■	▨	▨	■	■	Moderate	Sundarbans Reserved Forest (Bangladesh)
Certified forest area	□	□	□	□	□	□	■	Moderate	Upper St. John River (USA)

Particularly compatible with the protected area category
Not incompatible with the protected area category
Not particularly or never suitable for the protected area category

*IRBM = integrated river basin management, see text

Not all protected areas designated in whole or part to protect inland waters, including most Ramsar sites, have categories assigned. Additionally, many protected areas contributing to inland water ecosystem conservation have no Ramsar status. Consequently, it is presently not possible to assess globally which existing protected areas have inland water objectives, or how IUCN categories have been applied to them. Different types of inland water systems, with different degrees of intactness, may lend themselves more to some protected area categories than others: Table 12 makes some suggestions.

Table 12. Most appropriate protected area categories for different types of inland wetland ecosystems

Freshwater ecosystem type	IUCN category							Examples
	Ia	Ib	II	III	IV	V	VI	
River systems								
Entire catchments	■	■	■					Kakadu National Park (Australia)
Entire river/stream or substantial reaches						■	■	Fraser Heritage River (Canada)
Headwaters			■					Adirondack Forest Reserve (United States)
Middle and lower reaches		■						Doñana National Park (Spain)
Riparian zones			■				■	Douglas River/Daly River Esplanade Conservation Area (Australia)
Sections of river channels				■				Hippo Pool National Monument (Zambia)
Gorges				■				Fish River Canyon Conservation Area (Namibia)
Waterfalls				■				Iguacu National Park (Argentina/Brazil)
Wetlands and lakes								
Floodplain wetlands			■				■	Mamirauá Sustainable Development Reserve (Brazil)
Lakes			■	■				Lake Balaton (Hungary)
Portions of lakes					■	■		Rubondo Island National Park (Tanzania)
Inland deltas	■		■				■	Okavango Delta Wildlife Management Area (Botswana)
Coastal deltas			■				■	Danube Delta Biosphere Reserve (Romania)
Coastal wetlands			■				■	Doñana National Park (Spain)
Geothermal wetlands			■					Lake Bogoria (Kenya)
Springs	■		■					Ash Meadows National Wildlife Refuge (USA)
Alpine and tundra wetlands	■	■	■					Bitahai Wetland (China)
Freshwater swamps		■	■	■			■	Busanga Swamps (Zambia)
Peatland	■	■	■	■				Silver Flowe National Nature Reserve (UK)
Subterranean wetlands								
Karstic waters and caves	■			■				Mira Minde Polje and related Springs (Portugal)

Integrated protection of terrestrial and inland wetland systems

It is often difficult to identify an "inland water protected area" and the influence of a protected area on aquatic systems may have as much to do with its management objectives than its component habitats. Marine protected areas are easily identified by their location. Inland water systems, however, span the terrestrial landscape and occur in virtually all terrestrial protected areas. Certain protected areas, such as free-flowing rivers and many Ramsar sites, might clearly qualify as "inland water protected areas", but the designation of other sites can be ambiguous. Some have included both terrestrial and freshwater management goals from the outset, whereas others originally designated to protect terrestrial features have grown to incorporate freshwater objectives over time. South Africa's Kruger National Park is one example: originally designated to protect its large mammalian fauna, the riparian and riverine zones are estimated to support 50 percent of the park's biota and management now includes an estimated 30 percent inland water management focus.

Although some protected areas benefit the inland wetland systems within them, there are numerous other examples where this is not the case. In many instances, inland wetland ecosystems within protected areas have been deliberately altered to supply water and hydroelectricity, and even to facilitate wildlife viewing and other forms of recreation. Integration of inland wetland considerations into the management of *all* relevant protected areas is needed, including coastal MPAs. Management of terrestrial protected areas could better address inland waters, for example by:

- Protecting or restoring longitudinal and lateral connectivity of stream corridors (e.g., removing barriers, reconnecting rivers with floodplains, ensuring that roads and associated infrastructure within protected areas are not fragmenting stream systems);
- Protecting native fauna (e.g., prohibiting exotic fish stocking or overfishing);
- Protecting native flora – particularly in riparian zones which may be neglected in the broader protected area;
- Managing aquatic recreational activities (e.g., restricting motorized watercraft and discharge from boats);
- Aggressively protecting water quality (e.g., careful management of point-source discharges from recreational facilities);
- Protecting headwater flows so that downstream users can enjoy the benefits of ecosystem services;
- Protecting or restoring riparian buffers both within a park and along a park's border if a river demarcates the border (and extending PA boundaries where possible using appropriate inland wetland ecosystem criteria – e.g., using catchment boundaries, not river channels, to demarcate areas);
- Special protection for sacred springs or pools that have cultural significance.

In part because of continued ambiguity about whether or not an area is an "inland water protected area", separating out these components in recording processes such as the WDPA remains a challenge. Measuring and interpreting the size of many wetlands can be difficult, and in many cases wetlands vary greatly due to natural factors (e.g., seasonal flooding), and currently the WDPA has no provision for length measurements. Until inland water conservation is incorporated more

effectively into protected area management plans, and those management plans acknowledge processes and threats external to protected area boundaries, the geographic extent of inland water systems within protected areas tells us more about conservation potential than conservation reality.

Sacred natural sites

Sacred sites (including sacred natural sites and landscapes) that fit into national and international definitions of protected areas can potentially be recognised as legitimate components of protected area systems and can be attributed to any of the six IUCN protected area categories. At the same time, the cultural and spiritual values of protected areas should be better reflected in the whole range of categories, from which they are currently absent or insufficiently recognised.

Many protected areas contain sites of importance to one, and sometimes more than one faith or spiritual value systems, including both sacred natural sites and built monuments such as monasteries, temples, shrines and pilgrimage trails. Even in systems of protected areas in the most secularized countries of Europe, which were established using only ecological criteria, it is estimated that between 20–35 percent include significant cultural or spiritual values. There are countries and territories where all nature is sacred and protected areas can form smaller entities as part of larger sacred landscapes. Managers have to ensure that these spiritual values are protected alongside natural heritage. However, sacred sites are currently not effectively reflected in protected area designations and management plans, and existing policy and legal frameworks do not adequately support sacred (natural) sites. There is sound and widespread evidence that sacred natural sites have been providing effective biodiversity conservation, often for hundreds of years. Sacred sites may exist in more or less natural ecosystems, cultural landscapes or managed landscapes and when they occur in protected areas they need to be fully incorporated into management strategies in cooperation with the relevant faith and community groups. Some examples are given in Table 13.

Table 13.   Examples of sacred sites in IUCN categories

Ia	Strict nature reserve: protected area managed mainly for science		
	Sri Lanka	Yala National Park	Significant to Buddhists and Hindus and requiring high levels of protection for faith reasons.
	Russian Federation	Yuganskiy Kanthy	Significant to Christianity. The protected area has been created around Lake Numto – a Khanty and Nenets sacred place – in Beloyarsk region.
Ib	Wilderness area: protected area managed mainly for wilderness protection		
	Mongolia	Bogd Khan Mountain	The Mountain is significant to Buddhism and previously to shamanism. The Mountain has been officially designated as a sacred mountain by the state. Evidence exists of wilderness area declaration dating from 1294.
	Mongolia	Dornod Mongol	Significant to Buddhism. Vangiin Tsagaan Uul (White Mountain of Vangi) is a sacred Buddhist peak within the reserve.
II	National park: protected area managed mainly for ecosystem protection and recreation		
	Malawi	Nyika National Park	Large area containing four sacred sites, which local people can still use for rainmaking ceremonies.
	Japan	Kii Mountains National Parks and WHS	Several Shinto and Buddhist temples, sacred sites and pilgrimage trails for both faiths in continuous use for over one millennium.
	India	Great Himalayan National Park	Includes many places of religious importance for Hinduism.
III	Natural monument: protected area managed mainly for conservation of specific natural features		
	Cambodia	Phnom Prich Wildlife Sanctuary	A small area within the sanctuary is a sacred forest and therefore a natural monument (another example are the *Kaya* forests of Kenya).
	Russian Federation	Golden Mountains of Altai	Sacred to indigenous Altaians and many different faiths including Buddhist, Christian and Islamic.
	Greece	Mount Athos WHS peninsula	A stronghold of Orthodox Christianity including 20 monasteries contained within a monastic state and hundreds of smaller monastic settlements, hermitages and caves with over one millennium of continuous monastic activity.
	Spain	Montserrat Nature Reserve and Natural Park	Holy mountain containing ancient hermitages and a Christian monastery which have been a pilgrimage centre since the 14th century. Today it is the most heavily visited protected area of Spain.
IV	Habitat/species management area: protected area managed mainly for conservation through management intervention		
	Lebanon	Qadisha Valley and the Forests of the Cedars of God WHS	Sacred forest to the Christian Maronite Church, including a significant monastery, hermitages, and residence of religious authorities.
	Borneo	*tembawang* gardens	Some sacred sites will need continual intervention or even to be planted, such as the *tembawang* gardens that contain high levels of biodiversity.
	Sri Lanka	Peak Wilderness Park, (Sri Pada-Adams Peak)	Sacred natural site for Islam, Buddhism, Hinduism and Christianity, attracting many pilgrims of all these faiths.

Table 13.   Examples of sacred sites in IUCN categories (cont.)

V	Protected landscape/seascape: protected area managed mainly for landscape/seascape conservation and recreation		
	China	Xishuangbanna National Park	Landscape with several sacred sites (groves and mountains), which have long been managed by the community.
	Romania	Vanatori Neamt Natural Park	The spiritual heart of Romania, including 16 Christian monasteries, along with outstanding wildlife: European bison, brown bear and wolf populations.
VI	**Managed resource protected area: protected area managed mainly for the sustainable use of natural ecosystems**		
	Ecuador	Cayapas Mataje	Sustainable use area said to contain the world's tallest mangroves and known for important spirit dwellers that are worshipped by local people.
	USA	San Francisco Peaks National Forest	Sacred to over one dozen Native American tribes.
	Egypt	St Catherine Area WHS – Mt Sinai	Mount Sinai is sacred to Judaism, Christianity and Islam. The ancient monastery of St Catherine is a World Heritage site.

Where possible, the custodians of sacred sites should participate in their management. Traditional custodians of sacred sites should communicate and translate cultural and spiritual values of sacred sites to help to determine the management objectives. Sacred sites offer an excellent opportunity to engage in this dialogue and develop synergies that are environmentally sustainable and socially equitable.

Sacred sites and protected area categories

Whether or not particular sacred natural sites should be formally included in national protected area systems depends on the desires of the faith group and on whether or not the site's management objectives meet the IUCN definition of a protected area and the requirements of a particular category. This implies that the faith group recognises and agrees with the importance of maintaining biodiversity alongside the sacred values of the site.

Care needs to be taken to ensure that cultural and spiritual values do not jeopardise biodiversity values and that conversely the management of a protected area does not damage the site's sacred values. Integrating sacred sites, or more broadly, the perception of sacredness of nature, effectively into conservation plans is only possible when approached across ideological, physical and institutional borders. In short this is a process which integrates knowledge and wisdom with biodiversity conservation. Therefore, including sacred sites in all protected area categories builds on their intercultural and crosscutting values which, in turn, can produce equitable synergies between spiritual, cultural and natural diversity in support of more holistic conservation objectives.

Geodiversity

"Geodiversity is the variety of rocks, minerals, fossils, landforms, sediments and soils, together with the natural processes which form and alter them".

Many protected areas contain important geodiversity and some protected areas are designated primarily for their geodiversity values; in both cases maintenance of these values requires special consideration in management policies. Geodiversity is included under the term "nature conservation" in IUCN's definition of a protected area.

Geodiversity provides the foundations for life on Earth and for the diversity of natural habitats and landscapes. Many individual geological features and landforms have cultural or iconic values for humans, which influence the way that we view surrounding natural or semi-natural habitats. Geodiversity has also had a profound influence on many aspects of cultural landscapes, built environments and economic activities. Protection of geodiversity can be in response to a range of interests, including those associated with important fossil sites; reference sites for geoscience; spectacular features linked with tourism; and landforms that have particular cultural or spiritual values. Geodiversity can contribute to sustainable economic development through tourism associated with geological features. Understanding the functional links between geodiversity and biodiversity is particularly important for conservation management in dynamic environments, where natural processes (e.g., floods, erosion and deposition) maintain habitat diversity and ecological functions. This is explicit in the Ecosystem Approach and is fundamental at a time when many ecosystems face the impacts of climate change. Geodiversity is therefore a key consideration in sustainable management of the land, rivers and the coast. It requires integrated approaches to the management of the natural heritage, land and water at a landscape/ecosystem scale, based on understanding and working with natural processes and their likely responses to climate change.

Although category III provides an obvious focus for protection of specific geological features or landforms, geodiversity can be, and is, found protected in all IUCN categories and under all governance types. Some examples follow in Table 14.

Table 14. **Examples of geodiversity in different IUCN protected area categories**

Category	Example	Country
Ia		
Ib		
II	Grand Canyon National Park	United States
III	Jenolan Karst Conservation Reserve	Australia
IV		
V	Brecon Beacons National Park	UK
VI		

Although not definitive, Table 15 below gives some indication of when geodiversity values might match particular IUCN protected area categories

Table 15. **Indications of suitable IUCN protected area categories for different aspects of geodiversity**

Particular aspect of geodiversity under consideration	Category/categories suitable
Protection is aimed primarily at an individual feature of interest (natural monument such as a waterfall or cave) or a site of national or international value for geoscience.	Primarily category III
An assemblage of landforms (e.g., glaciated valley land system) and/or processes, or geological features.	Primarily categories Ia, Ib, II and V
The features have potential for interpretation and geotourism.	Primarily categories II and III
The geodiversity is itself a foundation for habitats and species (e.g., calcium-loving plants or species adapted to caves).	Primarily categories Ia, Ib, II, IV, V and VI
Geodiversity has important links with cultural landscapes (e.g., caves used as dwellings or landforms adapted to terraced agriculture).	Primarily category V also categories II and III
Geodiversity is the basis for sustainable management (activities associated with natural processes, such as cave tourism).	Primarily compatible with categories V and VI

Restoration and IUCN protected area categories

The IUCN protected area category is chosen primarily with respect to management objective, i.e., it relates to the *aims* of management rather than the current *status*, so that any category can be subject to restoration. However, in practice the category also usually infers something about the protected area status and active restoration is usually not suitable for every category of protected area. For example, categorization with respect to wilderness values (Ib) is not usually appropriate for an area that will require indefinite active management interventions to maintain these values. In some situations, restoration in a protected area can be a time-limited intervention to undo past damage while in other cases changes have been so profound that continued, long-term intervention will be needed: this is often true if some ecological components, such as important species, have disappeared. Some intervention, such as control of invasive species and in certain habitats and conditions prescribed burning, may be necessary in any category. The following advice describes the general situation but exceptions will occur:

- **Restoration through natural processes as a result of protection** (*mise en défens*): for instance restoration of old-growth forest through removal of logging or grazing pressure; recovery of fish stocks or coral reefs by restricting fishing; removal of trampling pressure in mountain plant communities – suitable for any category of protected area.
- **Restoration through time-limited interventions to undo past damage:** one or more interventions to restore damage; for example reintroduction of extirpated species; replanting to hasten forest regeneration; seedling selection; thinning; removal of invasive species – not usually suitable in strictly protected category Ia or Ib protected areas but usually suitable in other categories.[15]

---

[15] It is possible for a protected area to be re-categorized as a category Ia or Ib protected area if restoration is successful.

- **Restoration as a continual process for biodiversity conservation:** for instance artificial maintenance of water levels in a wetland in a watershed that has undergone major hydrological change; coppicing (regular cutting) of trees to maintain an important cultural forest; using domestic livestock grazing to maintain biodiversity values – generally suitable for categories IV–VI.

- **Restoration as a continual process for both natural resources and biodiversity:** for instance recovering productivity after soil erosion, providing resources for human well-being – suitable for categories V–VI.

Table 16.   Indicative guide to restoration in different IUCN categories

IUCN category						
Ia	Ib	II	III	IV	V	VI
Restoration through natural processes as a result of protection						
		Active, time-limited restoration				
				Continuous restoration for biodiversity		
					Continuous restoration for biodiversity and human needs	

In cases where general habitat destruction has advanced so far that protected areas themselves require substantial restoration, it may be sensible to wait and see how successful restoration projects are before assigning a category. The required degree of restoration and active management may increase in many protected areas under conditions of climate change.

7. International conservation initiatives

There are a number of parallel attempts to protect key habitats under the United Nations or regional agreements. Of particular relevance are the Convention on Biological Diversity, UNESCO natural World Heritage sites, UNESCO Man and the Biosphere reserves and Ramsar sites. The following section looks at how in particular Ramsar and World Heritage relate to the IUCN categories.

World Heritage Convention

World Heritage sites make up some of the most important cultural and natural places in the world recognised by the UNESCO World Heritage Convention and accorded particular protection by their host nations. They include monuments such as Angkor Wat in Cambodia or the Pyramids of Egypt, and also exceptional natural areas, such as Serengeti National Park in Tanzania or Canaima National Park in Venezuela. Governments nominate sites for possible inclusion on the World Heritage List, with recognition depending on a technical evaluation[16] followed by a review and final decision by World Heritage Committee members. Suitability is based in part on whether or not the site has Outstanding Universal Values (OUV), a term referring to the combination of those heritage values of a site that demonstrate how it is of global value and the requirement for a site to have integrity and effective management. IUCN is officially recognised in the text of the Convention as an Advisory Body for all natural and mixed natural-cultural sites. This involves carrying out technical evaluations of all applicant sites and also running monitoring missions as required for existing sites that may be under threat. Virtually all natural World Heritage sites are also protected areas. In the past, World Heritage sites were listed separately on the *UN List of Protected Areas* but this resulted in duplication, because many were also listed under their IUCN category.

What the World Heritage Convention requires from natural sites on the World Heritage List

The following notes aim to help governments considering the relationship of natural World Heritage sites to the IUCN protected area categories system. They do not cover cultural sites, most of which will not be in protected areas (or if they are will only be so by accident).

The relationship between World Heritage and protected areas in theory

The 2008 version of the World Heritage Convention's *Operational Guidelines* (OG) explains what is required under World Heritage (WH). It states that an area may be inscribed onto the list of WH only if the site meets the relevant World Heritage criteria and if strict conditions of integrity and conservation are met (paragraph 88), meaning that it must:

- Include all elements necessary to express the *Outstanding Universal Value* for which it is being nominated for inscription to the WH list;
- Be of adequate size to ensure the complete representation of the features and processes which convey the site's significance;
- Not suffer from adverse effects of development and/or neglect.

Potential WH sites are judged against several criteria, two of which (ecosystems and biodiversity) are particularly relevant to protected areas. Paragraphs 94–95 describe integrity for these two criteria:

- **Criterion ix (ecosystems):** the site "should have sufficient size and contain the necessary elements to demonstrate the key aspects of processes that are essential for the long term conservation of ecosystems and the biological diversity they contain".
- **Criterion x (biodiversity):** the site "should contain habitats for maintaining the most diverse fauna and flora characteristic of the bio-geographic province and ecosystems under consideration".

The OG acknowledges that "*no area is totally pristine and that all natural areas are in a dynamic state, and to some extent involve contact with people. Human activities, including those of traditional societies and local communities, often occur in natural areas. These activities may be consistent with the OUV of the area where they are ecologically sustainable*" (para. 90).

Finally, it includes a section entitled *Protection and Management* (para. 96–118), which outlines measures for the long-term conservation of areas nominated for WH consideration. Specifically, paragraph 97 states that: "*All properties inscribed on the World Heritage List must have adequate long-term legislative, regulatory, institutional and/or traditional protection and management to ensure their safeguarding. This protection should include adequately delineated boundaries.*" Paragraph 98 of the OG further adds that: "*Legislative and regulatory measures at national and local levels should assure the survival of the property and its protection against development and change that might negatively impact the outstanding universal value, or the integrity … of the property. States Parties should also assure the full and effective implementation of such measures*".

In regards to the relationship between nominated sites and existing protected areas, the OG state, in paragraph 102, that: "*The boundaries of the nominated property may coincide with one or more existing or proposed protected areas, such as national parks or nature reserves, biosphere reserves […]. While such established areas for protection may contain several management zones, only some of those zones may satisfy criteria for inscription*". This statement implies that some areas with legal protection might still not qualify for WH status, i.e., some forms of legal protection are not restrictive enough to satisfy the OG requirements.

Thus while the OG do not say that a site has to be a "protected area", or refer to IUCN protected area categories, it could be inferred that areas not under any particular protection regime

---

[16] All natural sites are evaluated by IUCN and all cultural sites are evaluated by ICOMOS – the International Council on Monuments and Sites.

should be excluded from WH sites (e.g., OG paragraphs 97 and 102): so natural World Heritage sites are expected to be managed in ways that are equivalent to being in a protected area, whether or not they are formally protected. This is the interpretation applied by IUCN in its advisory role.

The relationship between World Heritage sites and protected areas in practice

Having an effective management regime is a requirement for World Heritage listing and in practice this has meant that the vast majority of natural World Heritage Sites are protected areas. The UNEP-WCMC prepares data sheets for all proposed World Heritage sites and this explicitly lists the IUCN PA category under which the proposed site falls. There is thus a clear linkage between natural World Heritage sites and the categories system.

This situation has developed over time. In the earlier years of the Convention, some natural World Heritage sites included developments which would not be accepted today by the World Heritage Committee. As a result, some WH sites contain areas of incompatible uses large enough to be considered as

clearly defined zones within a WH site and not just minor "pre-existing" intrusions to an otherwise relatively undisturbed protected area. States Parties could in theory propose amendments to excise some of these areas from their older nominations. This is happening in a few cases although requires careful consideration on a case-by-case basis.[17]

Most existing and currently nominated WH sites correspond with existing protected area boundaries. Where large gaps separate protected areas that have similar and complementary values there is the potential to inscribe a serial nomination and such nominations are increasingly common (e.g., Discovery Coast Atlantic Forest Reserves in Brazil and Cape Floral Region Protected Areas in South Africa). The case studies outlined in Table 17 demonstrate how this tightening up has taken place over the last 25 years. New WH sites have gradually conformed more strictly to IUCN's definition of a protected area and areas not benefiting from a protection regime have increasingly been excluded. However some exceptions continue to occur (e.g., Peninsula Valdés in Argentina) and it is still not a requirement for a natural World Heritage site to be an official protected area if adequate protection and management can be provided by other means.

Table 17. Changing relationship between natural World Heritage sites and protected areas over time

Site name	WH criteria	IUCN cat.	Year inscr.	Discussion
Galapagos Islands ECUADOR	vii, viii, ix, x	II (land) IV (marine)	1978	Among the first batch of nominations ever submitted for inscription to the WH list, the terrestrial boundaries do not exclude the agricultural and settlement areas, resulting in a WH site that includes extensive cattle ranches and densely populated urban areas. The site was extended to include a marine protected area in 2001, which contains a mix of low-intensity multiple-use zones (diving, artisanal fishing).
Great Barrier Reef AUSTRALIA	vii, viii, ix, x	V	1981	A multiple-use zone, with a variety of permitted uses, from strict conservation to recreational including fishing. In its nomination evaluation report, IUCN suggested that the actual WH boundaries be limited to the fully protected core area (such comments not observed in the Galapagos nomination evaluation), but ended up recommending, in the same report, that the nomination as originally proposed be inscribed.
Lake Baikal RUSSIA	vii, viii, ix, x	Ia, II, IV	1996	This site consists of several distinct conservation management entities, along with non-conservation lands (e.g., coastal protection zones) of limited conservation value. A range of potentially incompatible uses occur, including commercial fishing, logging, agriculture, hunting and tourism. Several small settlements also occur in the site. Original recommendations for the WH site boundary had included a much vaster area, including major cities, but a smaller area with fewer conflicting uses was finally inscribed.

---

[17] For example, excision of ski resort areas from the existing World Heritage site is currently under consideration in the Pirin National Park in Bulgaria.

Table 17.   Changing relationship between natural World Heritage sites and protected areas over time (cont.)

Site name	WH criteria	IUCN cat.	Year inscr.	Discussion
East Rennell SOLOMON ISLANDS	ix	n/a	1998	Approximately 800 people of Polynesian origin reside in the site. Subsistence agriculture, fishing and hunting are carried out. The local people rely on forest products for most construction materials. The land is under customary ownership and a freshwater lake is regarded as common property. This was the first natural World Heritage site to be inscribed on the World Heritage list while under a customary management regime. In this case the WH Committee, on the recommendation of IUCN, noted that the customary management regime was sufficiently effective to ensure the protection of natural values.
Peninsula Valdès ARGENTINA	x	II, IV, VI	1999	A collection of seven distinct protected areas along with significant (e.g., >50 percent) proportion of private lands. Land owners are encouraged to collaborate through a joint management planning exercise, though are apparently not legally bound to do so. Current threats include land subdivision for coastal residential development. This appears to be an experiment in private land ownership within a natural WH site.
Discovery Coast Atlantic Forest Reserves BRAZIL	ix, x	Ia, II	1999	A series of eight distinct protected areas spread over 450 km² and nested within a one million ha biosphere reserve – interstitial lands are largely privately owned.
Cape Floral Region Protected Areas SOUTH AFRICA	ix, x	Ib, II, IV	2004	The inscription of this serial site is the result of a multi-year process through which the State Party's original nomination was not accepted due to the lack of a consolidated management regime for the collection of seven protected areas. As a result, a final nomination was submitted, meeting the technical requirements of IUCN, and inscribed by the WH Committee.
Sichuan Panda Reserves CHINA	x	n/a	2006	The original boundaries proposed by the State Party included towns, agricultural areas and public infrastructure works. Revisions of the original nomination took place over more than 10 years. IUCN requested the revision of the boundaries so that only core protected areas were included. The final boundaries reflect IUCN's request.

The relationship between World Heritage sites and IUCN protected area categories

It follows that if not all natural World Heritage sites are protected areas, not all will have IUCN categories. But in practice most are protected areas and most do have categories. Natural World Heritage sites occur in all the IUCN categories, but with a distinct bias towards the more strictly protected management objectives of category Ia, Ib and II. As at June 2008, there are 166 natural and 25 mixed World Heritage properties. Of these, 139 are inscribed under criteria ix and/or x (and thus focus on biodiversity/species issues), either exclusively, or in combination with the non-biodiversity criteria vii and viii and are considered as "biodiversity" natural heritage sites. Figure 4 illustrates the frequency of occurrence of a particular IUCN protected area category within natural WH sites.[18]

Over 70 percent of the World Heritage sites listed for biodiversity values contain (wholly or in part) a category II protected area. Some of these same sites may also contain protected areas of other categories (for instance, Te Wahipounamu in New Zealand is comprised of several different protected areas representing five different protected area categories). The chart shows that very few biodiversity WH sites contain category V and VI protected areas (these categories are represented in eight and six WH sites respectively, out of 128 sites for which the UNEP-WCMC database attributes a protected area category). Of these, only three (2 percent of all biodiversity sites) are comprised exclusively of a category V or VI protected area – being Australia's Great Barrier Reef (which is changing), Mauritania's Banc d'Arguin National Park (e.g., usually considered category II), and Tanzania's Ngorongoro Conservation Area. These are typically large sites: 348,700 km², 12,000 km² and 8,288 km² respectively.

---

[18]   Because a WH site may be composed of more than one PA, to which different categories are assigned, the numbers do not add up to 100 percent. Also, only 128 of the 139 biodiversity sites, and 38 out of the 47 non-biodiversity sites are attributed a PA category in the WDPA database.

Figure 4.   Frequency of IUCN PA categories occurrence in biodiversity and non-biodiversity natural WH sites

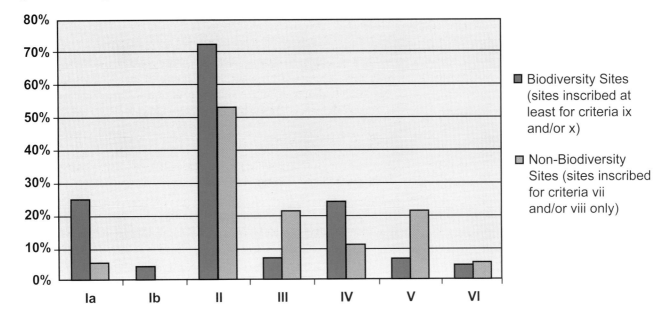

Conclusions

For governments considering nominating a natural World Heritage site:

- All natural World Heritage sites must have an effective management regime. This implies that such areas will be designated protected areas in virtually all cases.
- There is no rule to say that such sites have to be assigned an IUCN category, but again this is strongly encouraged and in fact all WCMC data sheets for proposed natural WH sites include an IUCN category(s) which corresponds to the proposed site. Most sites inscribed under criterion (ix) or (x) correspond to the IUCN category I or II; however there are exceptions and any category can be acceptable.

Ramsar Convention

The Ramsar Convention encourages Parties to designate and manage important wetlands in a way that does not change their ecological character. The 158 Contracting Parties (Governments) have committed themselves to the "wise use" of all wetlands on their territory (including rivers), conservation of "wetlands of international importance" (Ramsar sites), and international cooperation. Parties each commit to undertaking an inventory of their wetlands and preparing a "strategic framework for the Ramsar list" for the systematic and representative national designation and management of wetland habitat types. The Convention has many benefits for wetland conservation since it creates moral pressure for member governments to establish and manage wetland protected areas; sets standards, provides guidance, and facilitates collaboration on wise use; it has a triennial global reporting and monitoring system;

and encourages participation of NGOs, local communities and indigenous peoples.

While many of the Wetlands of International Importance (Ramsar Sites) also have other protection status (e.g., are protected areas under natural legislation, World Heritage sites or UNESCO biosphere reserves), there is no *obligation* for Ramsar sites to be legally protected areas **under national legislation**. Indeed, this sometimes helps to persuade governments to designate sites under Ramsar when they would be reluctant to make them national protected areas.

The protection afforded by the Convention is itself a legal support, but under soft law and not always so clearly articulated. For example, the *Criteria for Identifying Wetlands of International Importance* makes no reference to protection status. The *Information Sheet on Ramsar Wetlands* implies that protection status is not mandatory, with phrases such as: "If a reserve has been established". The *Ramsar Convention Manual* (2006) is explicit: "*Designating a wetland for the Ramsar List does not in itself require the site previously to have been declared a protected area*". In fact, listing under the Ramsar Convention, especially in the case of sites subject to intensive use by human communities – either to extract resources or to benefit from the natural functions of the wetland – can provide the necessary protection to ensure its long-term sustainability. This can best be achieved by preparing and implementing an appropriate management plan, with the active participation of all stakeholders.

As implied above, listing a wetland under the Ramsar Convention, especially in the case of sites subject to intensive use by human communities, should provide **the necessary protection** to ensure its long-term sustainability. Listing under Ramsar elevates the sites to a higher status, focuses more attention upon them, and should contribute to their long-term conservation and wise use – whether or not Ramsar status conveys additional legal protection in-country depends upon decisions of national and local governments. Human uses of wetlands are compatible with listing under Ramsar, provided that they meet the Ramsar concept of "wise use" (sustainable use) and do not lead to a negative change in ecological character.

The Ramsar Secretariat has sometimes viewed the Ramsar List as a set of "protected areas": for example the document *Emergency solutions seldom lead to sustainability* gives "an introduction to the concept of Wetlands of International Importance *as a network of protected areas*" (emphasis ours). Some Parties regard inclusion on the List as, in effect, meaning that the site becomes a protected area (whether or not it has an IUCN category), while others do not.

The categories system and Ramsar sites

In the original version of the management categories, biosphere reserves and World Heritage sites were identified as a category in their own right, yet Ramsar sites were not so identified. Subsequently, the 1994 guidelines did not treat any international designation as a separate category. It was agreed at the Ramsar Ninth Conference of Parties (Resolution IX.22) to include data about the IUCN category within the database of Ramsar sites. Out of the 84 sites designated since 1st January 2007, 37 (44 percent) include information on the IUCN category. Ramsar sites are nationally designated. The IUCN categories system is a means of classifying them on the basis of management objectives. Ramsar sites cut right across this approach because the very concept embodies the idea of a range of management objectives. On the other hand, some Ramsar sites often contain a series of management zones with differing management objectives, each of which may correspond to a category in the IUCN system. Some may consist of a number of different use categories.

The IUCN guidelines provide several ways in which the many different situations likely to be found within Ramsar sites can be reconciled with the categories system. Once it has been determined that the site meets the IUCN definition of a protected area, we recommend a two-stage approach:

- **Stage I:** *identify whether the whole Ramsar site should be classified under one, or more than one, category.*

To do this, it is necessary to establish which of three theoretical possibilities applies:

1. *There is only one management authority for the entire Ramsar site and, for legal purposes, the whole Ramsar site is classified by law as having one primary management objective.*

The area would be assigned to a single category.

While the guidelines require that the assignment be based on the primary purpose of management, they also recognise that management plans often contain management zones for a variety of purposes to take account of local conditions. In order to establish the appropriate category, at least three-quarters, and preferably more, must be managed for the primary purpose; and the management of the remaining area must not be in conflict with that primary purpose.

2. *There is one management authority responsible for two or more areas making up the Ramsar site, but each such area has separate, legally defined management objectives.*

The guidelines recognise this situation by acknowledging that "*protected areas of different categories are often contiguous, while sometimes one category 'nests' within another*". Thus many category V areas contain within them category I and IV areas: some will adjoin category II areas. Again, some category II areas contain category Ia and Ib areas.

In this case the separate parts of the Ramsar site will be categorized differently.

3. *There are two or more management authorities responsible for separate areas with different management objectives, which jointly make up the Ramsar site.*

Here, too, the correct interpretation of the guidelines would be to categorize these areas separately.

- **Stage 2:** *assignment of parts of the Ramsar site to individual categories.*

The categories system can be applied to a range of different legal and management situations which characterize Ramsar sites in different countries. This is entirely in line with the way in which the system is intended to be applied. IUCN states that protected areas should be established to meet objectives consistent with national, local or private goals and needs (or mixtures of these) and only then be labelled with an IUCN category according to the management objectives. These categories have been developed to facilitate communication and information, *not* to drive the system.

Benefits

The benefits of a system that can be applied internationally, in a transparent way, are significant. The principal advantage is that it allows global assessments of the existing Ramsar sites. Furthermore it facilitates development and further establishment of a Ramsar site system in which each country can maintain its individual Ramsar site network, yet be clearly part of the global framework of protected areas. It also allows the Ramsar site network to relate and contribute to the development of a globally comprehensive, adequate and representative system of protected areas.

It is intended to produce more detailed guidance on links between Ramsar sites and IUCN protected area categories.

Convention on Biological Diversity

At the seventh meeting of the Conference of the Parties (COP 7) to the CBD in 2004, 188 Parties agreed to a *Programme of Work on Protected Areas*, one of the most ambitious environmental strategies in history. The Programme aims, by 2010 (terrestrial) and 2012 (marine), to establish *"comprehensive, effectively managed and ecologically representative national and regional systems of protected areas"*. It has over 90 specific, time-limited target actions for member states and others.

Specifically, the Programme *"recognizes the value of a single international classification system for protected areas and the benefit of providing information that is comparable across countries and regions and therefore welcomes the on-going efforts of the IUCN WCPA to refine the IUCN system of categories and encourages Parties, other Governments and relevant organizations to assign protected area management categories to their protected areas, providing information consistent with the refined IUCN categories for reporting purposes"*.

The CBD has agreed its own definition of a protected area as a: *geographically defined area which is designated or regulated and managed to achieve specific conservation objectives*. There is tacit agreement between the CBD Secretariat and IUCN that the two definitions effectively mean the same thing. Significantly, the CBD Programme of Work explicitly recognises the IUCN protected area categories:

> Explore establishment of a harmonized system and time schedule for reporting on sites designated under the Convention on Wetlands, the World Heritage Convention, and UNESCO MAB programme, and other regional systems, as appropriate, taking into account the ongoing work of UNEP-WCMC on harmonization of reporting **and the IUCN protected area management categories system** for reporting purposes (*our emphasis*)

At the ninth CBD Conference of Parties, in 2008, support for the categories was reasserted and confirmed:

"9. Reaffirms paragraph 31 of decision VII/28, which recognizes the value of a single international classification system for protected areas and the benefit of providing information that is comparable across countries and regions and therefore welcomes the ongoing efforts of the IUCN World Commission on Protected Areas to refine the IUCN system of categories and encourages Parties, other Governments and relevant organizations to assign protected-area management categories to their protected areas, providing information consistent with the refined IUCN categories for reporting purposes".

There is therefore clear guidance from the CBD that countries should use the IUCN categories system in reporting progress on establishing and maintaining protected area systems.

8. Effectiveness of the IUCN categories

IUCN has always stressed that category is based on objective and is independent of effectiveness: that is if a protected area is failing to meet its objective this is not an excuse for shifting it to another category (but rather to increase management capacity). But many stakeholders are demanding a closer relationship between categories and effectiveness: the following section explores some options.

Assessment of management and the IUCN categories

Management effectiveness of protected areas has gained increasing attention as an essential element in maintenance of a successful protected area system, and *evaluation* or *assessment* of management is now seen to be a very useful tool in increasing effectiveness, by providing concise and practical information for managers and others. *Management effectiveness evaluation is defined as the assessment of how well protected areas are being managed* – primarily the extent to which they are protecting values and achieving goals and objectives. The term "management effectiveness" reflects three main "themes" in protected area management:

- design issues relating to both individual sites and protected area systems;
- adequacy and appropriateness of management systems and processes;
- delivery of protected area objectives including conservation of values.

Evaluation is now recognised as a component of responsive, proactive protected area management; both to help managers to make day-to-day decisions about allocation of time and resources and also increasingly as a stage in reporting progress on conservation in an international context, through conventions

and agreements such as the CBD. In the CBD's *Programme of Work on Protected Areas*, for example, signatory states have committed to develop systems of assessing management effectiveness and to report on 30 percent of their protected areas by 2010.

Evaluation of management effectiveness can:

- Enable and support an adaptive approach to management;
- Assist in effective resource allocation;
- Promote accountability and transparency;
- Involve the community and build support for protected areas.

The range of reasons for carrying out an evaluation combined with the great diversity of protected areas – with different values and objectives, cultural settings, management regimes and challenges – means that it is not practical to develop a single assessment tool. For this reason, IUCN-WCPA decided to develop a common framework (2nd edition, Hockings *et al.* 2006), which provides a consistent basis for designing assessment systems, gives guidance about what to assess and provides broad criteria for assessment. The process of assessment recommended by IUCN is summarised in Table 18 below. Based on this framework, a range of evaluation "tools" can be used to conduct evaluations at different scales and depths.

Table 18.   Elements of the WCPA framework for assessing management effectiveness of protected areas

	Design		Appropriateness/Adequacy		Delivery	
	Context	**Planning**	**Inputs**	**Process**	**Outputs**	**Outcomes**
Evaluation focus	Importance Threats Policy environment	Design and planning	Resources needed to manage	How management is conducted	Implementation of management programmes and actions	Extent to which objectives have been achieved
Criteria that are assessed	Values Threats Vulnerability Stakeholders National context	Legislation and policy System design Management planning	Adequacy of resources available for management	Suitability of management processes	Results of management actions	Effects of management in relation to objectives

A large number of systems for assessing management effectiveness have been developed over the past 10–15 years although many of these have been applied in only a few protected areas. More than 90 percent of site assessments have been undertaken using systems compatible with the IUCN-WCPA framework. This means that they share a common underlying approach and largely common criteria, although the indicators and assessment methods will vary. The systems can be broadly divided into two main types: (1) systems using mainly expert knowledge and (2) systems using data monitoring, stakeholder surveys and other quantitative or qualitative data sources. Some assessment systems combine both approaches to evaluation depending on the

aspect of management being assessed. The expert knowledge systems generally use a questionnaire approach asking people with detailed knowledge of the protected area and its management to rate various aspects of management or to nominate characteristics of the site such as the nature and significance of protected area values and threats. These assessments may be supported by a considerable knowledge base consisting of the results of monitoring and research carried out at the site. This approach to assessment is often applied when assessing management of large numbers of protected areas, often all of the protected areas in a country, as it is quicker and less resource-intensive than the monitoring approach.

Relationship between assessment and category assignment

Assessment can cover two different aspects of protected areas:

- Whether the objectives agreed for the protected area match the category being assigned. This becomes of more than academic interest if national policy or legislation links decision making (regarding e.g., funding, allowable land use, hunting rights etc.) to a category designation.
- Whether those objectives are being effectively delivered.

The first of these is basically an assessment of management intent. The purpose of such assessments is not to evaluate the effectiveness of management but to clarify the expressed and implemented objectives for management. Such an approach has been developed by IUCN-WCPA in Europe and has been used to "certify" that a protected area has been assigned to the correct protected area category (according to legislation and governing regulations) and whether the site is being managed in accordance with management objectives relevant to that category. As yet, there is no written methodology and the system is under development. It focuses particularly on the first two elements in the WCPA framework – context and planning – and hardly at all on the last two of outputs and outcomes.

The second looks more deeply at whether these objectives have been delivered in practice. These objectives are normally specified at national level in relevant legislation or other governance system (e.g., traditional authority for community conserved areas) which provides overall direction for management of the site. For example, designation as a category II protected area means that the area should be managed primarily for biodiversity conservation with no, or very limited, extractive use of resources. In some cases, managers may have difficulty in managing the site in strict accordance with these objectives. It has been assumed that the results of assessments of management effectiveness should not be used as a basis for allocating or changing the category to which a protected area is assigned. So, for example, the appropriate response to an evaluation of management effectiveness that reveals a failure to control illegal resource exploitation in a category II protected area is not to change the site to category V (which allows for a level of sustainable resource use) but rather, to seek to adapt management to achieve more effectively the legally specified management objectives.

In future, IUCN will be investigating the demand for more rigorous assessment of effectiveness within the context of the categories system and looking at practical implications.

Appendix. Typology and glossary

Users will be reading these guidelines line by line, trying to make difficult judgements, frequently working in something other than their first language. So the guidelines must be as clear as possible but precision is made more difficult by the fact that many words used in ecology and conservation remain vaguely defined and subject to multiple interpretation. The glossary in Table 19 is offered to give clarity and should be used in conjunction with the definition and descriptions of categories that follow. Sources used have wherever possible drawn on previous IUCN definitions or those of the CBD and should thus be familiar to governments and others using the categories.

Table 19. Definition of terms used in the guidelines

Term	Definition	Source and notes
Agrobiodiversity	Includes wild plants closely related to crops (crop wild relatives), cultivated plants (landraces) and livestock varieties. Agrobiodiversity can be an objective of protected areas for crop wild relatives, traditional and threatened landraces, particularly those reliant on traditional cultural practices; and/or traditional and threatened livestock races, especially if they are reliant on traditional cultural management systems that are compatible with "wild biodiversity".	**Source:** Amend, T., J. Brown, A. Kothari, A. Phillips and S. Stolton (Eds). 2008. *Protected Landscapes and Agrobiodiversity Values.* Volume 1 in the series Values of Protected Landscapes and Seascapes. Heidelberg: Kasparek Verlag, on behalf of IUCN and GTZ.
Biological diversity	The variability among living organisms from all sources including, *inter alia*, terrestrial, marine and other aquatic ecosystems and the ecological complexes of which they are part; this includes diversity within species, between species and of ecosystems.	**Source:** CBD, Article 2. Use of Terms http://www.cbd.int/convention/articles.shtml?a=cbd–02 **Translations:** text available on CBD website in Arabic, Chinese, English, French, Russian, Spanish.
Biome	A major portion of the living environment of a particular region (such as a fir forest or grassland), characterized by its distinctive vegetation and maintained largely by local climatic conditions.	**Source:** From the Biodiversity Glossary of the CBD Communication, Education and Public Awareness (CEPA) Toolkit: http://www.cbd.int/cepa/toolkit/2008/cepa/index.htm
Buffer zone	Areas between core protected areas and the surrounding landscape or seascape which protect the network from potentially damaging external influences and which are essentially transitional areas.	**Source:** Bennett, G. and K.J. Mulongoy. 2006. *Review of experience with ecological networks, corridors and buffer zones.* Technical Series no. 23. Montreal: Secretariat of the CBD (SCBD).
Community Conserved Area	Natural and modified ecosystems, including significant biodiversity, ecological services and cultural values, voluntarily conserved by indigenous peoples and local and mobile communities through customary laws or other effective means.	**Source:** Borrini-Feyerabend, G., A. Kothari and G. Oviedo. 2004. *Indigenous and Local Communities and Protected Areas: Towards Equity and Enhanced Conservation.* Best Practice Protected Area Guidelines Series No. 11. Gland and Cambridge: IUCN.
Corridor	Way to maintain vital ecological or environmental connectivity by maintaining physical linkages between core areas.	**Source:** Bennett, G. and K.J. Mulongoy. 2006. *Review of experience with ecological networks, corridors and buffer zones.* Technical Series no. 23. Montreal: SCBD.
Ecosystem	A dynamic complex of plant, animal and micro-organism communities and their non-living environment interacting as a functional unit.	**Source:** CBD, Article 2. Use of Terms http://www.cbd.int/convention/articles.shtml?a=cbd–02 **Translations:** Arabic, Chinese, English, French, Russian, Spanish.

Table 19. Definition of terms used in the guidelines (cont.)

Term	Definition	Source and notes
Ecosystem services	The benefits people obtain from ecosystems. These include provisioning services such as food and water; regulating services such as regulation of floods, drought, land degradation, and disease; supporting services such as soil formation and nutrient cycling; and cultural services such as recreational, spiritual, religious and other non-material benefits.	**Source:** Hassan, R., R. Scholes and N. Ash (Eds). 2005. *Ecosystems and Human Well-Being: Current State and Trends: Findings of the Condition and Trends Working Group v. 1 (Millennium Ecosystem Assessment)*. Washington DC: Island Press. Definitions in: Chapter 1: MA Conceptual Framework.
Framework	A high-level structure which lays down a common purpose and direction for plans and programmes.	**Source:** The CBD Communication, Education and Public Awareness (CEPA) Toolkit: http://www.cbd.int/cepa/toolkit/2008/cepa/index.htm This definition is from the CEPA Glossary; which is an updated version of a communication glossary developed by the IUCN CEC Product Group on Corporate Communication, edited by Frits Hesselink in 2003.
Geodiversity	The diversity of minerals, rocks (whether "solid" or "drift"), fossils, landforms, sediments and soils, together with the natural processes that constitute the topography, landscape and the underlying structure of the Earth.	**Source:** McKirdy, A., J. Gordon and R. Crofts. 2007. *Land of Mountain and Flood: the geology and landforms of Scotland*. Edinburgh: Birlinn.
Governance	In the context of protected areas, governance has been defined as: "*the interactions among structures, processes and traditions that determine how power is exercised, how decisions are taken on issues of public concern, and how citizens or other stakeholders have their say*". Governance arrangements are expressed through legal and policy frameworks, strategies, and management plans; they include the organizational arrangements for following up on policies and plans and monitoring performance. Governance covers the rules of decision making, including who gets access to information and participates in the decision-making process, as well as the decisions themselves.	**Source:** Borrini-Feyerabend, G., A. Kothari and G. Oviedo. 2004. *Indigenous and Local Communities and Protected Areas: Towards Equity and Enhanced Conservation*. Best Practice Protected Area Guidelines Series No. 11. Gland and Cambridge: IUCN.
Governance quality	How well a protected area is being governed – the extent to which it is responding to the principles and criteria of "good governance" identified and chosen by the relevant peoples, communities and governments (part of their sense of morality, cultural identity and pride) and generally linked to the principles espoused by international agencies and conventions.	**Source:** Borrini-Feyerabend, G. 2004. "Governance of protected areas, participation and equity", pp. 100–105 in Secretariat of the Convention on Biological Diversity, *Biodiversity Issues for Consideration in the Planning, Establishment and Management of Protected Areas and Networks*. Technical Series no. 15. Montreal: SCBD.
Governance type	Governance types are defined on the basis of "who holds management authority and responsibility and can be held accountable" for a specific protected area.	**Source:** Borrini-Feyerabend, G. 2004. "Governance of protected areas, participation and equity", pp. 100–105 in Secretariat of the Convention on Biological Diversity, *Biodiversity Issues for Consideration in the Planning, Establishment and Management of Protected Areas and Networks*. Technical Series no. 15. Montreal: SCBD.

Table 19.   Definition of terms used in the guidelines (cont.)

Term	Definition	Source and notes
In-situ conservation	The conservation of ecosystems and natural habitats and the maintenance and recovery of viable populations of species in their natural surroundings and, in the case of domesticated or cultivated species, in the surroundings where they have developed their distinctive properties.	**Source:** CBD, Article 2. Use of Terms http://www.cbd.int/convention/articles.shtml?a=cbd–02 **Translations:** Arabic, Chinese, English, French, Russian, Spanish.
Indigenous and tribal people	(a) Tribal peoples in independent countries whose social, cultural and economic conditions distinguish them from other sections of the national community, and whose status is regulated wholly or partially by their own customs or traditions or by special laws or regulations; (b) Peoples in independent countries who are regarded as indigenous on account of their descent from the populations which inhabited the country, or a geographical region to which the country belongs, at the time of conquest or colonization or the establishment of present State boundaries and who, irrespective of their legal status, retain some or all of their own social, economic, cultural and political institutions.	**Source:** Definition applied to the International Labour Organization (ILO) Convention (No. 169) concerning Indigenous and Tribal Peoples in Independent Countries. Indigenous peoples also stress that there is a degree of self-definition in determining what makes up a specific indigenous or tribal people.
Management effectiveness	How well a protected area is being managed – primarily the extent to which it is protecting values and achieving goals and objectives.	**Source:** Hockings, M., S. Stolton, F. Leverington, N. Dudley and J. Courrau. 2006. *Evaluating Effectiveness: A framework for assessing management effectiveness of protected areas. 2nd edition.* Best Practice Protected Area Guidelines Series No. 14. Gland and Cambridge: IUCN. **Translations:** Forthcoming in French and in Spanish.
Sacred site	An area of special spiritual significance to peoples and communities.	
Sacred natural site	Areas of land or water having special spiritual significance to peoples and communities.	**Source:** Wild, R. and C. McLeod. 2008. *Sacred Natural Sites: Guidelines for Protected Area Managers.* Best Practice Protected Area Guidelines Series No. 16. Gland and Cambridge: IUCN.
Shared governance protected area	Government-designated protected area where decision-making power, responsibility and account ability are shared between governmental agencies and other stakeholders, in particular the indigenous peoples and local and mobile communities that depend on that area culturally and/or for their livelihoods.	**Source:** Borrini-Feyerabend, G., A. Kothari and G. Oviedo. 2004. *Indigenous and Local Communities and Protected Areas: Towards Equity and Enhanced Conservation.* Best Practice Protected Area Guidelines Series No. 11. Gland and Cambridge: IUCN.
Stakeholder	Those people or organizations which are vital to the success or failure of an organization or project to reach its goals. The primary stakeholders are (a.) those needed for permission, approval and financial support and (b.) those who are directly affected by the activities of the organization or project. Secondary stakeholders are those who are indirectly affected. Tertiary stakeholders are those who are not affected or involved, but who can influence opinions either for or against.	**Source:** The CBD Communication, Education and Public Awareness (CEPA) Toolkit: http://www.cbd.int/cepa/toolkit/2008/cepa/index.htm This definition is from the CEPA Glossary; which is an updated version of a communication glossary developed by the IUCN CEC Product Group on Corporate Communication, edited by Frits Hesselink in 2003.

Table 19.   Definition of terms used in the guidelines (cont.)

Term	Definition	Source and notes
Sustainable use	The use of components of biological diversity in a way and at a rate that does not lead to the long-term decline of biological diversity, thereby maintaining its potential to meet the needs and aspirations of present and future generations. (This definition from the CBD is specific to sustainable use as it relates to biodiversity).	**Source:** CBD, Article 2. Use of Terms http://www.cbd.int/convention/articles. shtml?a=cbd–02 **Translations:** Arabic, Chinese, English, French, Russian, Spanish.

References

Bishop, K., N. Dudley, A. Phillips and S. Stolton. 2004. *Speaking a Common Language – the uses and performance of the IUCN System of Management Categories for Protected Areas.* Cardiff University, IUCN and UNEP/WCMC.

Borrini-Feyerabend, G., A. Kothari and G. Oviedo. 2004. *Indigenous and Local Communities and Protected Areas: Towards equity and enhanced conservation.* Best Practice Protected Area Guidelines Series No. 11. Gland and Cambridge: IUCN.

Bridgewater, P., A. Phillips, M. Green and B. Amos. 1996. *Biosphere Reserves and the IUCN System of Protected Area Management Categories.* Canberra: Australian Nature Conservation Agency.

Brockman, C.F. 1962. "Supplement to the Report to the Committee on Nomenclature". In: Adams, A.B. (Ed.) *First World Conference on National Parks.* Washington, DC: National Park Service.

CBD. Undated. http://www.cbd.int/programmes/cross-cutting/ecosystem/default.shtml. Accessed 24 August 2007.

Chape, S., S. Blyth, L. Fish, P. Fox and M. Spalding. (Eds). 2003. *2003 United Nations List of Protected Areas.* Gland and Cambridge: IUCN and UNEP-WCMC.

Davey, A.G. 1998. *National System Planning for Protected Areas.* Best Practice Protected Area Guidelines Series No. 1. Gland and Cambridge: IUCN.

Day, J. 2002. "Zoning: Lessons from the Great Barrier Marine Park". *Ocean and Coastal Management* 45: 139–156.

Dillon, B. 2004. "The Use of the Categories in National and International Legislation and Policy". *PARKS* 14(3): 15–22.

Dudley, N., L. Higgins-Zogib and S. Mansourian. 2006. *Beyond Belief: Linking faiths and protected area networks to support biodiversity conservation.* Gland and Bath: WWF International and Alliance on Religions and Conservation.

Dudley, N. and J. Parrish. 2006. *Closing the Gap: Creating ecologically representative protected area systems.* Technical Series no. 24. Montreal: Secretariat of the CBD.

Dudley, N. and A. Phillips. 2006. *Forests and Protected Areas: Guidance on the use of the IUCN protected area management categories.* Best Practice Protected Area Guidelines Series No. 12. Gland and Cambridge: IUCN.

Dudley, N. and S. Stolton. 2003. "Ecological and socio-economic benefits of protected areas in dealing with climate change". In: Hansen, L.J., J.L. Biringer and J.R. Hoffman (Eds) *Buying Time: A user's guide to building resistance and resilience to climate change in natural systems*, pp. 217–233. Washington, DC: WWF US.

Eidsvik, H. 1990. A Framework for Classifying Terrestrial and Marine Protected Areas. Based on the Work of the CNPPA Task Force on Classification, IUCN/CNPPA. Unpublished.

Elliott, H.B. (Ed). 1974. *Second World Conference on National Parks, Proceedings.* Morges: IUCN.

EUROPARC and IUCN. 1999. *Guidelines for Protected Area Management Categories – Interpretation and Application in Europe.* Grafenau: EUROPARC.

Graham, J., B. Amos and T. Plumptre. 2003. *Principles for Good Governance in the 21st Century.* Policy Brief Number 15. Ottawa: Institute on Governance.

Hockings, M., S. Stolton, F. Leverington, N. Dudley and J. Courrau. 2006. *Evaluating Effectiveness: A framework for assessing management effectiveness of protected areas.* 2nd edition. Gland and Cambridge: IUCN.

Holdaway, E. Undated. Making the Connection between Land and Sea: The place for coastal protected landscapes in the marine environment. Wadebridge and Bangor: EUROPARC Atlantic Isles and the Countryside Council for Wales.

Holdgate, M. 1999. *The Green Web.* London: Earthscan.

IUCN. 1974. *Classification and Use of Protected Natural and Cultural Areas.* IUCN Occasional Paper No. 4. Morges: IUCN.

IUCN. 1978. *Categories, Objectives and Criteria: Final Report of the Committee and Criteria of the CNPPA/IUCN.* Morges: IUCN.

IUCN/WCMC. 1994. *Guidelines for Protected Area Management Categories.* Gland and Cambridge: IUCN.

IUCN. 2004. *PARKS* 14. (includes 10 papers).

Kelleher, G. 2002. *Guidelines for Marine Protected Areas.* Best Practice Protected Area Guidelines Series No. 3. Gland and Cambridge: IUCN.

Palumbi, S.R. 2001. "The ecology of marine protected areas". In: Bertness, M.D, S.M. Gaines and M.E. Hixon (Eds). *Marine Community Ecology*, pp.509–530. Sunderland, MA: Sinauer Associates.

Phillips, A. 2002. *Management Guidelines for IUCN Category V Protected Areas: Protected Landscapes/Seascapes.* Best Practice Protected Area Guidelines Series No. 9. Gland and Cambridge: IUCN.

Phillips, A. 2007. "A short history of the international system of protected area management categories". Paper prepared for the WCPA Task Force on protected area categories.

Sandwith. T., C. Shine, L. Hamilton and D. Sheppard. 2001. *Transboundary Protected Areas for Peace and Cooperation.* Best Practice Protected Area Guidelines Series No. 7. Gland and Cambridge: IUCN.

Stolton, S., N. Dudley and J. Randall. 2008. *Natural Security: Protected areas and hazard mitigation*. The Arguments for Protection Series. Gland: WWF International.

Sulu, R., R. Cumming, L. Wantiez, L. Kumar, A. Mulipola, M. Lober, S. Sauni, T. Poulasi and K. Pakoa. 2002. "Status of Coral Reefs in the Southwest Pacific Region to 2002: Fiji, Nauru, New Caledonia, Samoa, Solomon Islands, Tuvalu and Vanuatu". In: Wilkinson, C.R. (Ed.) *Status of Coral Reefs of the World 2002*. Townsville, Queensland: Australian Institute of Marine Science.